I dedicate this book to all the patients with prostate problems that I have looked after, and to all our amazingly generous supporters

**Roger Kirby**

KENSINGTON PALACE
LONDON W8 4PU

**From HRH The Duchess of Gloucester, GCVO**

As Patron of Prostate Cancer UK, I am delighted to welcome the fourth edition of Professor Roger Kirby's excellent book. This 'easy to understand' book provides one of the most comprehensive reads available today on prostate diseases, and congratulations are due to Professor Kirby on his ability to write in this manner on such a complex subject.

We have learnt that time is of the essence. I cannot emphasize enough that the sooner a problem, or even a suspicion of something not being quite right, is dealt with, the greater the chance of recovery. The statistics continue to encourage a return to a full bill of health when a prostate problem is confronted and corrected without delay.

**Her Royal Highness
The Duchess of Gloucester, GCVO**

# CONTENTS

| | |
|---|---|
| **Introduction** | 1 |
| **Prostate health and awareness** | 3 |
| Function of the prostate | 4 |
| Common diseases involving the prostate | 4 |
| How you can help yourself | 5 |
| Seeing your doctor | 9 |
| | |
| **The PSA test** | 13 |
| Basis of the PSA test | 13 |
| Issues surrounding the PSA test | 15 |
| It's not always cancer | 19 |
| Your choice | 19 |
| | |
| **A raised or rising PSA: what happens next?** | 21 |
| Finding a good urologist | 21 |
| Tests | 22 |
| If cancer is suspected | 25 |
| Grading and staging tests | 26 |
| Partin's tables | 31 |
| Receiving bad news | 32 |
| | |
| **Prostate cancer: what it is and what causes it** | 35 |
| Stages of cancer | 36 |
| Why do some men get prostate cancer and others do not? | 38 |
| Can prostate cancer be prevented? | 39 |
| | |
| **Treatment options for prostate cancer that has not spread beyond the gland** | 43 |
| Active surveillance | 43 |
| Watchful waiting | 44 |
| Radiotherapy | 45 |
| Radical prostatectomy | 49 |
| Surgery or radiotherapy? | 55 |
| The long-term picture | 55 |

## CONTENTS

### If prostate cancer has spread or recurs after treatment    **63**
    Locally advanced disease    63
    Metastatic disease    66
    Recurrence    70

### Prostate cancer: the future    **77**
    Chemoprevention    77
    Genetics and targeted screening    77
    Better diagnosis    77
    New treatments    78

### BPH: its symptoms, diagnosis and treatment    **81**
    Why do some men suffer more than others?    81
    How is BPH diagnosed?    83
    Treatment    87
    Prevention    97

### Prostatitis: the painful prostate    **99**
    Risk factors    101
    Tests    101
    Treatment    102
    Prostatic abscess    105
    Preventing prostatitis    105
    Pain but no inflammation: prostatodynia    105

### Medications commonly used to treat prostate disorders    **107**
    Alpha-blockers    107
    5-alpha-reductase inhibitors    107
    LHRH analogues    108
    LHRH antagonist    108
    Anti-androgens    109
    Anticholinergics and anti-spasmodics    109
    Vasopressin analogues    110
    Other drugs    110

# CONTENTS

**Further information and support**     **113**
General health and lifestyle     113
Prostate disorders     114
Treatment     115
Continence     115
Sexuality     115
Practical support     116
Support groups     116

**Some medical terms explained**     **119**

**Index**     **129**

# Introduction

At the beginning of the 20th century, the average life expectancy for a man in Europe or the USA was a mere 49 years. As diseases of the prostate typically affect men beyond middle age, the likelihood of a man in the early 1900s suffering from one of these conditions was rather slim. Nowadays, however, life expectancy for men extends well into the 70s, and this increased longevity has been accompanied by a rising tide of prostate disease with resultant effects on quality of life. Over the next 20 years, life expectancy is predicted to rise still further, to 80 and beyond. What we are witnessing at the moment is, then, simply the tip of the iceberg. The number of men with a prostate problem is set to more than double by the year 2020.

Recently there has been a surge in public interest in the prostate, largely as the result of a spate of media attention. Scarcely a week goes by without a newspaper or television feature on this aspect of men's health. Prominent personalities, including Bob Monkhouse, Sir Harry Secombe and the former England rugby union star Andy Ripley, have also spoken openly about their prostate problems before, sadly, passing away as a result of the disease.

This increasing focus can only be good news, as men with prostate disease can increase their chance of cure with a little knowledge and timely action. Men, and their partners and family who love and support them, need to be aware of the symptoms and signs of prostate problems, and the possibilities of a simple examination and a blood and/or urine test. The fourth edition of this book contains the essence of the things that you and those closest to you need to know to respond appropriately to the threat of these widespread and troublesome diseases. Use this information to obtain the best treatment for you – you owe it to yourself and to those who are dear to you.

# Prostate health & awareness

The prostate is a walnut-sized gland that is present only in men. It is located deep in the pelvis, at the exit of the bladder, and surrounds the tube known as the urethra (through which urine flows from the bladder to the outside of the body). Tiny at birth and throughout childhood, the prostate enlarges after puberty, stimulated by rising levels of the male hormone testosterone secreted by the testes, to a volume of around 20cc. Although the prostate is small compared with other organs, it looms ever larger as a potential source of disease and disability once a man passes middle age.

The prostate is subdivided into three zones: central, transition and peripheral, which are shown in the diagram overleaf. The peripheral zone is located at the back of the prostate and is the part most susceptible to both prostate cancer and prostatitis.

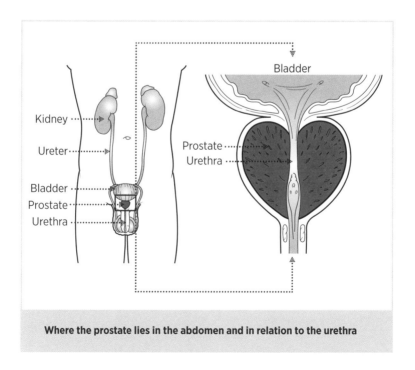

Where the prostate lies in the abdomen and in relation to the urethra

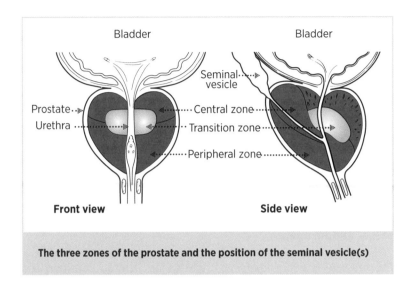

The three zones of the prostate and the position of the seminal vesicle(s)

The third and most common prostate problem – benign prostatic hyperplasia (BPH) – develops in the transition zone, which lies in the middle of the gland and surrounds the urethra.

## Function of the prostate

The prostate gland manufactures an important liquefying component of semen. Sperm are produced in the testicles and then stored just behind the prostate in the seminal vesicles. Here the sperm are in a jelly-like medium. At orgasm and ejaculation, the prostate and seminal vesicles contract, mixing their respective contents. The fluid in the prostate contains large amounts of a substance known as prostate-specific antigen (PSA), which liquefies the gelatinous sperm mixture, allowing the sperm to move freely in search of an ovum to fertilize.

## Common diseases involving the prostate

Because the prostate surrounds the urethra, any disease of the gland is likely to cause disturbances in urinary flow, and in the frequency and efficiency with which the bladder is emptied. There are three common diseases that may affect the prostate, and consequently urinary flow and frequency of urination:

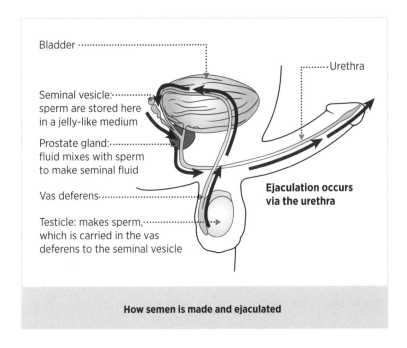

How semen is made and ejaculated

- BPH, which results in frequent urination and a reduced urinary stream, and affects almost 50% of men beyond middle age
- Prostate cancer, which is now the most common malignancy in men, with a 1 in 8 lifetime risk, rising to 1 in 4 for men of Afro-Caribbean descent:
- Prostatitis, an inflammatory disease affecting about 1 in 15 mainly younger and middle-aged men, which is characterized by symptoms of pain and discomfort around the anus, scrotum and the area in between (the perineum).

## How you can help yourself

Men's attitude towards their health has traditionally been 'if it ain't broke, don't fix it'. These days, the thinking should be more along the lines of 'if you look after your body (and particularly your heart and your prostate), it's less likely to go wrong'.

## Lifestyle

Adopting a healthier lifestyle, in terms of both diet and exercise, is obviously desirable. By staying slim and fit, you are more likely to remain healthy and, in particular, you will reduce your chances of developing cardiovascular disease and diabetes. This may also benefit your prostate, as there is some evidence that overweight men and those with higher cholesterol levels are more susceptible to prostate cancer. BPH is also more common in obese men, and surgery in obese individuals is more risky. In addition, prostatitis will often resolve as general health improves. In prostate disease, as in so many other areas, prevention is better than cure. I often advise my patients to use the mnemonic shown below to get to a better 'PLACE'.

## Diet

A healthy diet is essential for good health, and the best way to achieve this is to eat moderate quantities of a wide variety of the right foods. Try to have:

- less fat, particularly saturated fats, which are found in fatty meats and dairy products, and 'trans fats', which are found mostly in margarine and processed baked goods
- more fish and chicken (but not the skin), and less red meat
- less sugar and salt
- plenty of fresh fruit and vegetables – try and eat at least five portions and aim for nine portions a day (see page 115)
- foods high in fibre, such as wholewheat bread and grains
- moderate amounts of tea, coffee and alcohol

---

**Lifestyle**

A helpful mnemonic for healthy lifestyle. Get to a better PLACE:

**P**ortion control
**L**ose the booze
**A**xe the snacks
**C**ut the carbs
**E**xercise every day

- antioxidant-rich foods, such as blueberries, strawberries and broccoli.

## Selenium
Until recently, selenium was thought to reduce the incidence of all cancers and prostate cancer in particular, based on study results. However, a large study designed to specifically look at the effects of selenium and vitamin E, taken singly or in combination, has now disproved any protective effect, so it is no longer recommended.

## Zinc
Zinc deficiency is unusual, but may be responsible for some prostate problems, as well as impotence. It is therefore important to include good sources of zinc in your diet: for example, meat, fish, wholegrains and legumes, such as peas and beans.

## Vitamin D
Prostate cancer is more common in northerly latitudes. Sunlight increases the body's vitamin D level, which is believed to be protective against prostate cancer. A vitamin D supplement can be taken but as yet there is no definite proof that taking a supplement reduces the risk of prostate cancer.

## Antioxidants
Antioxidants are thought to protect the body's cells against cancer-causing substances. The main antioxidants, which are found in fruit and vegetables, are vitamins A, C and E, and lycopene, which is found in tomatoes. The two most effective antioxidants are vitamin E (*see Selenium*) and lycopene. Research has shown that those people who eat a lot of tomatoes and tomato products have a lower risk of certain cancers, particularly prostate cancer. Tomatoes are more effective when they have been processed or cooked, because heating with a little oil helps to release the lycopene from the tomato skin and makes it easier to absorb. Lycopenes can be purchased as a supplement, for example 'Lyco-mato' from your local chemist.

## Cranberry juice
Urinary tract infections are more common in men with an enlarged prostate gland. While such infections are not life-threatening or dangerous, they can take a considerable financial and social toll on those affected. Although more research is needed, drinking one or two

glasses of cranberry juice a day does seem to ward off urinary tract infections in some individuals. A word of caution: if you are taking the blood-thinning drug warfarin, you should avoid cranberry juice as it can interfere with the effects of the drug. Blueberry juice is thought to have similar properties and also contains antioxidants. Highly coloured fruits such as strawberries and raspberries are also recommended by some.

### Saw palmetto
The herb saw palmetto (*Serenoa repens*) is a plant native to the southeast of the USA. It has been shown to inhibit the action of 5-alpha-reductase (see page 90), growth factors and inflammatory substances responsible for the common symptoms of BPH (see page 83). A study reported in the New England Journal of Medicine called into doubt its efficacy; however, some patients swear by it, so more research is needed.

### Exercise
Regular exercise is essential for good health and plays an important role in reducing the risk of developing many diseases.

A recent study in America found that men who undertook 3 hours of vigorous exercise per week were 70% less likely to develop or die from prostate cancer. Exercise also helps us to maintain a healthy body weight, particularly when combined with a healthy diet. The prominent 'beer belly' that is so characteristic of overweight men is strongly linked to an increased risk of heart attack and stroke. It therefore makes sense to try to do some moderate physical exercise for at least 30 minutes five times or more a week. Remember that vigorous exercise not only burns off calories as you do it, but also increases the metabolic rate for up to 12 hours after the exertion. Consequently, twice-daily exercise is especially good if you are actively trying to lose weight.

### Smoking
Those readers, or their relatives and friends, who smoke are strongly exhorted to give it up! Although smoking does not cause prostate disease directly, it may result in bladder and kidney cancer and, because it damages blood vessels, it may also result in reduced erections and sexual dysfunction.

### Other 'therapies'
A host of other complementary therapies are claimed to protect against prostate cancer, but sound evidence for their safety and effectiveness is sparse. This includes soya products, green tea, apricot kernels, pomegranates and St John's wort. Cancer patients, in particular, are very vulnerable to hype about so-called 'alternative' therapies, many of which have no clinical or scientific basis. In some cases they may even be harmful, as they can contain toxic substances or may interact with medicines prescribed by your doctor. It is therefore important that you seek advice from your doctor before embarking on any alternative therapy. Much more research is needed to provide better evidence that various foods, plant extracts and supplements are truly as safe and effective as the manufacturers and retailers would have us believe.

## Seeing your doctor
Each of the three major prostatic diseases eloquently illustrates the 'stitch in time' principle. A 'window of curability' exists for prostate cancer, but once this is closed, neither surgery nor radiotherapy is likely, ultimately, to be successful. With BPH, several studies have confirmed that there is a level of secondary damage to the bladder caused by obstruction after which complete recovery becomes less likely. And if prostatitis becomes chronic, then repeated and prolonged courses of treatment are often needed for the relapses that occur.

### When should you go?
Regular prostate checks allow prostate disease to be detected at a stage when it can generally be resolved, while preventive strategies may reduce the risks of disease developing in the first place. It makes good sense to combine these with regular, more general health checks to diagnose other potentially dangerous conditions such as high blood pressure, raised cholesterol and diabetes. In a way, then, the prostate provides the key to more general men's health.

It is generally advisable for men over 50 to have an annual health check, which should include an assessment of the prostate, including a PSA test. While a one-off PSA check provides a certain amount of information, regular checks are more informative because they show the rate of the rise in PSA (sometimes called the 'PSA kinetics').

A sudden rise in PSA is rather like a flashing light on the dashboard of your car: it tells you that something is amiss, which, if responded to appropriately, will help to prevent eventual breakdown. You may have to ask for a PSA test as not all doctors will offer it routinely. Write down and keep the result as it's always a good thing to 'know your numbers'.

### Specific symptoms to prompt a visit

Problems with urinating are the most common symptoms of prostate disease. You should visit your doctor if you regularly experience one of the following:

- a weak, sometimes intermittent flow of urine
- difficulty starting to urinate
- a need to urinate frequently
- a need to urinate urgently (you do not feel able to put it off)
- having to go to the toilet several times during the night
- a feeling that your bladder is not completely empty after you have finished urinating
- pain or burning when passing urine
- blood in your urine (this is particularly important).

## Important

**See your doctor regularly and promptly**
- If men wait for symptoms, they may wait too long
- Men with prostate disease can be cured by a little knowledge and timely action
- A 'window of curability' exists for prostate cancer
- If cancer is to be identified at a stage when it is still curable, then it should be detected before the PSA rises much above 10 ng/mL (see next chapter)
- In both benign prostatic hyperplasia (BPH) and prostatitis, earlier treatment results in better outcomes

# The PSA test

PSA (prostate-specific antigen) is a protein-like substance that occurs in abundance in the fluid within the prostate. Testing blood samples to determine the amount of PSA (a 'PSA test') is still central to the early detection and selection of the most effective treatment for prostate cancer. Monitoring a man's PSA level is also extremely helpful once therapy has been started, as it can indicate how effectively treatment is working. If the PSA is rising in spite of treatment, second-line therapies such as abiraterone, enzalutamide or chemotherapy with, for example, docetaxel may be in order.

## Basis of the PSA test

Cancer can be defined as the uncontrolled division of cells. Prostate cancer develops from the lining cells of the tiny glands within the prostate whose function is to manufacture PSA. Not surprisingly therefore, prostate cancer cells nearly all continue to secrete PSA. As the cancer grows, PSA levels tend to rise. Moreover, as the pre-cancer stage, prostatic intraepithelial neoplasia (PIN, see page 36) evolves into invasive prostate cancer, the membrane surrounding the glands within the prostate may start to break down in small areas. As a consequence, the fluid in the prostate and the PSA it contains start to leak out. The PSA finds its way into the blood and so the amount of PSA in the blood starts to increase. Progressively worsening damage to the prostate makes it more leaky which, in turn, results in higher PSA levels in the blood.

A normal PSA level (in a man with no prostate problems) is sometimes accepted as being below 4 ng/mL (4 nanograms per millilitre), but this rises with age so that in men over 70 a cut-off of 6.5 ng/mL is accepted (see table overleaf). There is nothing magical about a cut-off value, however. Recent studies have shown that many men with a PSA below 4 ng/mL may, in fact, harbour small cancers. In younger men, especially, it is the rate of the rise in PSA rather than its absolute value that may be important. Current research suggests that a rise of more than 0.75 ng/mL per year may indicate the need for further investigation.

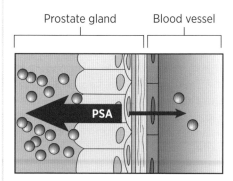

**Normal**

Cells in the prostate are healthy and organized in a tight pattern. Only a small amount of PSA leaks out of the prostate and gets into the bloodstream

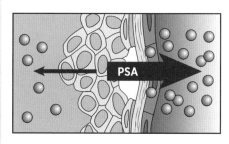

**With prostate cancer**

Now the cells are disorganized and the layers between the prostate and blood vessel become disrupted. More PSA can leak into the blood vessel as a result

The growth of cancer cells in the prostate disrupts the structure and organization of the tissue. PSA inside the prostate is able to leak into the nearby blood vessels more readily than it does in a healthy prostate. As a result, the amount of PSA in the blood increases, which is why measurement of PSA in a blood sample can help to diagnose prostate cancer

## INCREASE IN ACCEPTED PSA CUT-OFF WITH AGE

| Age | PSA cut-off |
| --- | --- |
| 40-49 years | 2.5 ng/mL |
| 50-59 years | 3.5 ng/mL |
| 60-69 years | 4.5 ng/mL |
| Over 70 years | 6.5 ng/mL |

PSA in the bloodstream is either free or bound to one of two proteins – antichymotrypsin and alpha macroglobulin. For reasons that are still not clear, in men with prostate cancer the amount of unbound or 'free' PSA is reduced. As a consequence, a reduction in the percentage of free PSA is also an early warning sign for prostate cancer. The cut-off point is usually taken as 18%; values above this suggest benign prostate enlargement, while values less than 18% increase the probability of prostate cancer being present.

## Issues surrounding the PSA test

When doctors and journalists talk about screening for prostate cancer, they are usually referring to the potential to test every man's PSA level at fixed intervals of time (like the smear test for women), from the age of around 50 onwards. If the test is so useful, why is it not used in this way? There are several points that have to be considered, and the pros and cons of the PSA test are summarized in the table opposite.

### Overdetection of clinically insignificant cancers

As prostate cancer occurs mainly in men beyond middle age, it is perfectly possible that a small cancer might never grow sufficiently large to cause problems during a man's lifetime. The anxiety caused by a 'positive' (high) PSA result might reduce the man's quality of life by causing unnecessary worry, whereas if he remained ignorant of his condition, his life would be unaffected. However, fears of overdiagnosis and overtreatment have lessened as doctors increasingly use 'active surveillance' as a treatment strategy for smaller low-risk cancers.

### 'False-positive' results

An elevated level of PSA in the blood does not necessarily indicate cancer. Indeed, the average PSA level rises with age and any disease of the prostate – particularly BPH, but also prostatitis – can result in an elevated PSA. A high PSA value, or one that increases over time, may prompt a doctor to request a biopsy, which involves taking samples of tissue from the prostate. However, scientific studies have shown that when samples of prostate tissue are examined under the microscope, only one man among four with a PSA value between 4 and 10 ng/mL will be found to have cancer (so three of the four will not have cancer even though their PSA levels are raised).

## THE PROS AND CONS OF PSA TESTING

**Pros**
- Allows early detection of potentially curable prostate cancer
- Permits the doctor to estimate prostate size in a patient with BPH
- Helps the doctor predict response to certain drugs
- Allows the doctor to estimate how advanced the cancer is at diagnosis
- Can be used to monitor men at increased risk of prostate cancer, such as those with a family history
- Can help the doctor estimate the patient's risk of developing prostate cancer in the future
- A negative result is reassuring
- Sequential values provide extra information about cancer risk
- Helpful for monitoring response to treatment

**Cons**
- Clinically insignificant cancers may be detected causing needless worry and further medical procedures for the patient
- Men without cancer may have a false-positive result (particularly those with borderline PSA values)
- A false-negative result may provide unwarranted reassurance
- There are cost implications – not only regarding the PSA test, but of biopsy and treatment options if the biopsy is positive
- Those undergoing biopsy are exposed to the risks of bleeding and infection

Using a higher cut-off value (say PSA above 10 ng/mL), the probability that a subsequent biopsy will confirm prostate cancer rises to more than 50% (every other patient). Of course, the problem with using a higher cut-off to determine who should undergo a biopsy is that as the cut-off value increases, so does the 'false-negative' rate. (A false-negative result is a PSA test result below the cut-off value, but the man has prostate cancer; this is illustrated in the diagram opposite. Early prostate cancer can be present even when the PSA value is below 4 ng/mL.) Also, if cancer is to be identified at a stage when it is still reliably curable, then it should be detected before the PSA rises much above 10 ng/mL. There has been great interest in measuring the rate of PSA change over time. Although the information on this subject is only provisional, as already mentioned, it seems that men whose PSA rises by more than 0.75 ng/mL per year are at higher risk of harbouring the more aggressive form of prostate cancer as opposed to a less aggressive type of tumour, which in fact carries little risk of spread (aggressive tumours are often described as 'tigers' and the more innocuous tumours as 'pussycats'). In order to detect the rate of PSA change, regular (usually yearly) blood tests are required. These can usefully be combined with cholesterol, lipid and blood sugar measurements, provided that fasting blood samples are obtained. Ask your doctor for the results, and make sure you keep a record.

### Anxiety before the results become available

The speed with which you get your test results depends on where you have your test. It can take anything from a couple of hours to several weeks – obviously those waiting at the longer end of the scale have more time to become anxious. Ask your doctor about the usual waiting time for results. Bear in mind, however, that minor fluctuations are common. Ask your doctor to explain any variations.

Over-the-counter PSA tests, which will allow self-testing, are available in chemist shops. Like the whole PSA issue, these home tests are rather controversial; it is best not to rely on them.

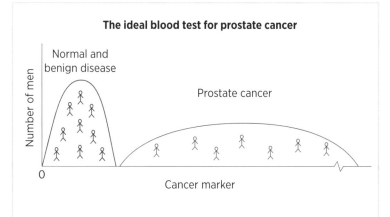

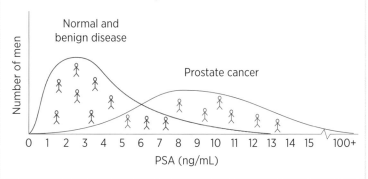

An 'ideal' blood test for prostate cancer should clearly distinguish men who have cancer from those who do not. Although PSA measurement is a useful test, it is not perfect – if a cut-off of 4 ng/mL is used, there is an overlap between patients with BPH or other diseases and those with prostate cancer. Inevitably, this results in some men worrying unnecessarily that they may have cancer

## It's not always cancer

It is worth re-emphasizing that a PSA level that is higher than normal does not necessarily mean that you actually have prostate cancer. Conversely, a normal PSA value does not conclusively exclude the presence of the disease. Both BPH and prostatitis can result in elevated PSA levels in the blood, and your doctor will cross-check your PSA result with your symptoms, the result of a digital rectal examination (see pages 22 and 23) and probably the results from an MRI scan and a biopsy to make the diagnosis. If you have a raised PSA, but a negative result on biopsy, your doctor will probably monitor your PSA level over time. Depending on further results, he may suggest that you have another biopsy at a later date. The value of sequential PSA testing lies in its ability to provide a baseline. A sudden or progressive rise above this level may act as an early warning of either prostate cancer development or another disease process within the gland. Urinary infection, for example, or sudden retention of urine requiring a catheter, can both cause the PSA level in the blood to rise sharply.

## Your choice

In the end, it is up to you whether you have the PSA test or not, based on your informed choice, and whether you continue with annual or biannual checks. But make sure you base your decision on reliable information, not the latest newspaper, radio or TV article, or some unsubstantiated internet site, and discuss it with your partner. The Department of Health has recently changed its policy on PSA testing and has agreed that men should be entitled to this test once they have had enough information to make an informed choice (see www.cancerscreening.nhs.uk/prostate/).

In summary, the PSA test is not perfect, and much work is currently being done to come up with something better. For the time being, however, it is the best we have, and if used intelligently can be a valuable early indicator of problems, benign and malignant, that are developing within the prostate. If caught early, these problems can nearly always be dealt with effectively and safely.

# A raised or rising PSA: what happens next?

## Finding a good urologist

If your GP finds that you have a raised or rising PSA level (usually above 4 ng/mL, but in younger men above 2.5 ng/mL), or a reduced percentage of free PSA (less than 18%) you will probably be referred to a urologist – a specialist in disorders affecting the kidney, bladder and prostate in men (and the urinary tract in women). It is important that you feel comfortable with, and confident in, your urologist. You should understand his explanations of procedures and options, and he should be prepared to discuss fully anything that concerns you or your partner. In this day and age, do not simply accept that the 'doctor knows best' – it is your health and peace of mind at stake here, so make sure that you have had all your questions answered before you leave the consultation room. If you or your family are not happy with your urologist, go back to your GP and discuss the matter with him and ask for a second opinion.

Alternatively, you may want to find your own specialist on a private basis. If this is the case, the first thing to do is to check your health insurance, if you have it. Some companies will not cover your expenses unless you have been referred by your GP. Also, you (or your insurers) may want to check the prices of treatment at an early stage. The clinic should provide a price list for you. If you are not happy with your service, talk to the clinic manager or the urologist directly – you are a prospective customer and they will be unlikely to want to 'lose' you. If you are still not happy, go back to your GP and discuss the issue.

If you find it difficult to voice your concerns face-to-face or if you feel that you might forget some things, write a letter or list so that you can make sure that all your points are answered. It may also be useful to take your partner or a friend with you to the consultation, as two people will often pick up more information than one. Taking a look at some of the websites listed from page 115 onwards may also be helpful, and these days you can find out about the doctor who is treating you, and the hospital or clinic in which he works, from the internet. In addition, most hospitals and clinics give out information sheets which you should study carefully.

## Tests

Some measurements will be taken to assess your general health. For example, height, weight and abdominal girth may be measured, as obesity is a risk factor for prostate cancer, diabetes and heart disease. Your blood pressure is also likely to be checked, as early identification of hypertension is important to prevent the development of complications. For similar reasons, blood tests for sugar and cholesterol (see opposite) may be carried out.

Your urologist may also repeat tests that your GP has already carried out, such as the PSA and percentage free PSA measurements. He may want to check your situation for himself; for some tests it is advisable that the samples are always sent to the same laboratory for analysis if you are being monitored over a period of time.

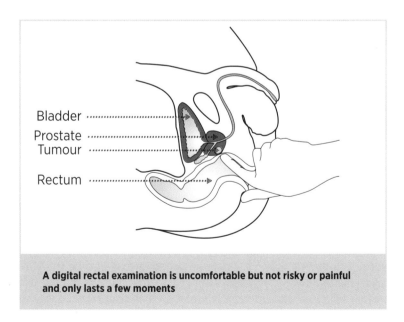

Bladder
Prostate
Tumour

Rectum

**A digital rectal examination is uncomfortable but not risky or painful and only lasts a few moments**

## Blood tests

The basis of the PSA test was described earlier on pages 13 – 17. Other tests, such as your fasting blood sugar and cholesterol level, may be used as indicators of your general health and to rule out diabetes or an abnormal lipid profile (e.g. raised HbA1c or cholesterol). They can also be used to estimate the risk of future problems, such as heart disease or stroke and as a guide for the need for preventative treatment.

## Urine tests

You may be asked to provide a urine specimen. This will be checked for bacteria, as you may have a urinary tract infection. It may also be tested for the presence of blood. Blood can be found in the urine if prostate cancer has spread into the urethra, which runs through the prostate, so its presence is a clue to the urologist about the nature of your problem. Other important causes of blood in the urine include bladder stones and bladder cancer, so this is a finding that should not be ignored. Urine may also be routinely tested for sugar in order to detect diabetes.

## Physical examination and digital rectal examination

The urologist may examine you in general, but will almost certainly perform a digital rectal examination. Undeniably, it is an uncomfortable experience and one that some men dread, but the discomfort is actually only mild. Urologists perform this day in, day out, but if that does not reassure you, just keep thinking about the consequences of ignoring your condition. A few moments of minor discomfort are surely worthwhile to establish a diagnosis.

Your urologist will put on a glove and apply some lubricant jelly to his finger. He will tell you which position to adopt – probably one where you lie on your side with your legs pulled up towards your chest. He will then gently insert his finger into your rectum, passing through the sphincter muscle that keeps the anus closed. He will then feel your prostate, noting its size, shape, firmness and how its surface feels – an enlarged but soft prostate suggests benign enlargement of the gland, while a firm nodule may indicate cancer. The examination is not painful, just uncomfortable. Try to relax until it is over – it literally only lasts a few moments.

| | Not at all | Less than 1 time in 5 | Less than half the time | About half the time | More than half the time | Almost always | Patient score |
|---|---|---|---|---|---|---|---|
| **1. Incomplete emptying** Over the past month, how often have you had a sensation of not emptying your bladder completely after you finished urinating? | 0 | 1 | 2 | 3 | 4 | 5 | |
| **2. Frequency** Over the past month, how often have you had to urinate again less than two hours after you finished urinating? | 0 | 1 | 2 | 3 | 4 | 5 | |
| **3. Intermittency** Over the past month, how often have you found you stopped and started again several times when you urinated? | 0 | 1 | 2 | 3 | 4 | 5 | |
| **4. Urgency** Over the past month, how often have you found it difficult to postpone urination? | 0 | 1 | 2 | 3 | 4 | 5 | |
| **5. Weak stream** Over the past month, how often have you had a weak urinary stream? | 0 | 1 | 2 | 3 | 4 | 5 | |
| **6. Straining** Over the past month, how often have you had to push or strain to begin urination? | 0 | 1 | 2 | 3 | 4 | 5 | |
| **7. Nocturia** Over the past month, how many times did you most typically get up to urinate from the time you went to bed at night until the time you got up in the morning? | 0 | 1 | 2 | 3 | 4 | 5+ | |
| **Total score** | | | | | | | |

| | Delighted | Pleased | Mostly satisfied | Mixed | Mostly dissatisfied | Unhappy | Terrible |
|---|---|---|---|---|---|---|---|
| **Quality of life due to urinary symptoms** If you were to spend the rest of your life with your urinary condition the way it is now, how would you feel about that? | 0 | 1 | 2 | 3 | 4 | 5 | 6 |

**A sample questionnaire on urinary symptoms**

### Urination questionnaire

Prostate cancer may be affecting your ability to empty your bladder or you may have BPH that is affecting your urine flow. In order to investigate your symptoms in a meaningful way, your urologist will probably give you a questionnaire to fill in (see opposite page). You may be asked to fill it in while you are in the clinic or you may be able to take it home with you to complete and return at your convenience.

## If cancer is suspected

If cancer is suspected, your urologist will first need to check whether you do in fact have cancer by requesting an MRI scan and/or performing a biopsy, which involves taking some tiny samples from the prostate while you have either a local or general anaesthetic. If you have, he will then need to determine how aggressive it is and how far it has progressed. You may hear a reference to the grade and stage of your cancer. These are important in selecting the best treatment option for you.

### Grade

The grade is a measure of how aggressive the cancer is. The cancer cells in the prostate start out looking very similar to normal prostate cells, but start to change their appearance and de-differentiate (i.e. become more aggressive) as the cancer progresses (see overleaf). Grading is a means of assessing this process in a standardized way, and is performed in a laboratory by specialist pathologists.

The standard grading system is the Gleason score. From post-mortem findings, we know that low-grade 'latent' prostate cancer is probably quite common among men aged over 40; these small tumours often grow very slowly, and so many men will never develop symptoms during their natural lifespan. However, a cancer that progresses more quickly (that is, a more aggressive one) will show a less 'differentiated' pattern, and will be graded higher. The cancerous areas in the prostate may vary and have different grades, so the grades of the two most prominent areas are added together to give a Gleason score (for example, 3 + 4); the maximum is 10 (5 + 5). This figure then gives your doctor an idea of how quickly your cancer is likely to progress and therefore helps him or her advise you about treatment.

# RESULTS

**GLEASON SCORE AND THE RISK OF PROSTATE CANCER PROGRESSING**

| GLEASON SCORE | RISK |
| --- | --- |
| 4 – 6 | Low |
| 7 | Medium |
| 8 – 10 | High |

### Stage

The cancer can also be classified according to how far it has spread, that is its 'stage'. The tumour–nodes–metastases (TNM) system is commonly used, and involves the doctor assessing how far your cancer (tumour) has spread in and around the prostate, whether it has spread to the nearby lymph nodes (nodes) and then whether it has spread (metastasized) to the distant lymph nodes and bones. Knowing the stage of your cancer helps you, your family and your urologist to decide on the most appropriate course of action.

## Grading and staging tests

PSA measurement and digital rectal examination are both important for staging, but you will almost certainly have to undergo some further tests.

### Ultrasound

Ultrasound may be used to assess the size and texture of the prostate; the specific technique is called transrectal ultrasonography (or TRUS for short). A lubricated probe is inserted into the rectum, where it passes high-frequency sound waves through the prostate. Computer analysis of the echoes, which vary according to the density of the tissues the waves are passing through, produces an image of the prostate that can then be seen on a screen. Ultrasound is a relatively simple and safe procedure that is not too uncomfortable, but without a biopsy it cannot be used to tell definitively whether or not cancer is present.

## Ultrasound-guided biopsy

Ultrasound-guided biopsy is used to obtain tiny samples of tissue from your prostate that can then be sent to the pathology laboratory for analysis under a microscope. The pathologist can check whether cancer is present and, if it is, grade it. You will probably be recommended for biopsy, an outpatient procedure, on the basis of your PSA level. Increasingly, doctors are requesting an MRI beforehand to help target the biopsies accurately.

Using ultrasound for guidance, a probe with a fine needle attachment is inserted into the back passage until it reaches the prostate (shown in the diagram below). The test is not too painful, but you may feel a

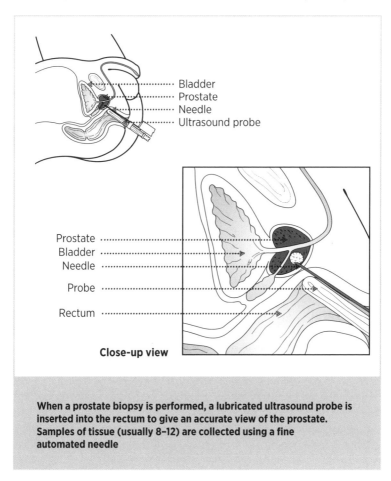

When a prostate biopsy is performed, a lubricated ultrasound probe is inserted into the rectum to give an accurate view of the prostate. Samples of tissue (usually 8–12) are collected using a fine automated needle

sharp needle prick with each biopsy, even though a local anaesthetic has been used, as 8–12 or sometimes more tissue samples are taken. The results should be available within a couple of weeks.

You will be given antibiotics (tablets or an injection) for 24 hours before and/or immediately before the procedure, and will be told to continue taking the prescribed antibiotic tablets for several days afterwards. For several weeks after the procedure, you may notice blood in your urine, semen and/or bowel motions. This is quite normal, but if you have any worries, consult your doctor. Urinary infections and septicaemia occur in 2–5% of men as a consequence of the biopsy – if you feel a burning sensation on urination, notice that your urine is cloudy and/or smelly, find that you have to urinate more frequently than normal and/or you develop a temperature, have shaking attacks and feel generally unwell, you must contact your doctor urgently. He will probably prescribe more antibiotics or even admit you to hospital for intravenous antibiotic treatment. E. coli bacteria are usually responsible for these problems.

## Transperineal biopsy

A variation on the standard biopsy is the template transperineal biopsy. This involves inserting a fine needle through the skin between the scrotum and the anus many times in order to obtain tissue samples from the prostate for testing. The procedure is usually carried out under a light general anaesthestic, which enables many more samples to be taken than would be tolerated using the standard method. It also reduces the rate of infection, though it may result in some difficulty passing urine for several days after the procedure. Some urologists recommend a period of catheterisation (insertion of a tube to drain the bladder) afterwards to avoid this.

It's important to appreciate, though, that biopsies of the prostate are only tiny samples of the whole gland, so small cancers may sometimes be missed. If the PSA continues to rise in spite of a negative biopsy result, a further set of biopsies may be required. Several studies have shown that prostate cancer is unlikely to be present in a man who has had three consecutive negative biopsies; BPH is likely to have triggered the PSA rise in this case.

Patients often ask whether having a biopsy will cause the cancer to spread. There is absolutely no evidence of this. The ability to spread (to metastasize) to other parts of the body, such as the bones, depends on the characteristics of the cancer cells themselves and tends to occur quite late in the disease.

## Bone scans

Bone scans are a means of checking whether the cancer has spread (metastasized) around the body to the bones. Three hours before you have the scan, a mixture containing radioactive particles (radionuclides) will be injected into your arm. The particles then spread around your body; their pattern, which shows up on the scanner, can reveal 'hot spots', which are dark areas of abnormal blood flow – a sign that cancer may be present.

'Hot spots' can also be the result of other diseases, such as arthritis in the joints and spine, so further testing may be necessary to clarify the cause of an abnormal scan. Do not be concerned about the use of radiation here – the amount is so low that the risk to your health is negligible. In general, men with a PSA less than 10 hardly ever have a bone scan that confirms the spread of the cancer.

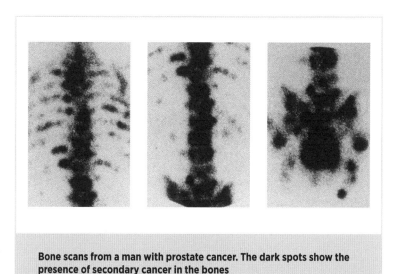

**Bone scans from a man with prostate cancer. The dark spots show the presence of secondary cancer in the bones**

## MRI

MRI, or magnetic resonance imaging, is a technique whereby a strong magnetic field and radio signals are used to examine sequential cross-sections of the body. The images that result are highly detailed – the radiologist and urologist can use them to assess the location and extent of the cancer in the prostate and to check whether any secondary tumours have formed in other regions. The procedure is completely painless, but some people find being in the scanner rather noisy and claustrophobic. The results should be available within a few days. If you have recently had a biopsy, the MRI will often be delayed for 4–6 weeks to allow the biopsy reaction to settle. If you have any metal implants, such as a pacemaker or coronary artery stents, it is usually not possible to perform an MRI scan, so a CT scan will probably be arranged as an alternative. Increasingly, MRI scans are being done before the biopsy to help the doctor decide which particular area of the prostate to target with the biopsy needle. The improved technology of 3Tesla MRI imaging with gadolinium enhancement allows more accurate visualisation of the cancer. If it is completely normal you may be able to avoid a biopsy.

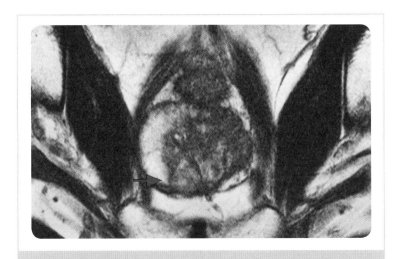

An MRI scan showing cancer (arrowed) in the prostate

### CT scanning

CT, or computed tomography, is similar to MRI in that the technique produces images of sequential slices through the body, but it uses X-rays to build up the images. CT scanning is not as accurate as MRI in terms of looking at the prostate, but is much less claustrophobic. Occasionally, CT scanning is used to guide biopsy needles to obtain tissue samples from enlarged lymph nodes or other soft tissues. It is also used when planning radiotherapy to target the prostate and the cancer within it.

### Choline PET scanning

A new technology known as choline PET scanning is increasingly being used to detect the spread of prostate cancer. Metastases in the bone and soft tissues such as lymph nodes and lungs may be revealed.

### Why scans are not always necessary

Although you might think that every man who has been diagnosed with prostate cancer requires scanning, in fact, in men with a PSA below 10 ng/mL, the chances of a positive bone scan are so low that it is often not recommended. Many patients feel reassured to know that their scans are clear, but bone scans can give false-positive results and MRI scans can also be misleading as they cannot reliably detect microscopic spread outside the prostate.

## Partin's tables

Although the tests described above seem very 'high tech' and sophisticated, unfortunately they do not always give a very precise answer to the question 'has the cancer spread beyond the prostate?' In fact the so-called 'Partin's tables', which compare the findings of the rectal examination, the PSA level and the Gleason score, are still often the best way of estimating the risk of spread beyond the prostate capsule.

These tables were developed by Dr Alan Partin, now the Professor of Urology at the renowned Johns Hopkins University Hospital, Baltimore, USA. He has shown that the smaller the cancer feels on rectal examination, and the lower the PSA level and Gleason score, the greater the likelihood that the cancer can be completely removed by surgery. These tables can therefore be useful in helping the doctor, patient and family decide together on the best treatment option.

## Receiving bad news

Of course, being told that you have prostate cancer will come as a major shock. In an instant your optimistic prospects for the future are transformed. The blow can be lessened, however, if the news is broken sensitively and sympathetically, in the presence of your partner or a close friend, by a caring and informed professional who gives you as much time as you need. It has been said, wisely, that if a doctor breaks bad news kindly, the patient will never forget him. But if done badly, the patient will never forgive him.

Often nowadays, the consultation with the doctor who delivers the news about the biopsy results is followed immediately by an interview with a urology nurse specialist. This is a move that is supported and endorsed by Prostate Cancer UK. This specialist nurse will help to answer any further questions you might have and provide written material about the disease and its treatment. This can be invaluable, because many patients retain only a fraction of the information they are given after the shocking news of a cancer diagnosis. The specialist nurse will also provide details of sources of support from charities, such as Marie Curie, Cancer Research UK and Macmillan Cancer Support, as well as details of patient support groups. (You can find details of sources of further information and support at the end of the book).

Although the news may not be what you had hoped for, remember that the outlook for men with prostate cancer nowadays is often quite good and, with all the current research effort and the development of new treatment options, is improving all the time.

# Prostate cancer: what it is and what causes it

Prostate cancer develops as a result of a progressive series of faults occurring in the genes that control cell growth in the prostate. These faults can be inherited or develop as a result of damage to the DNA, the material that controls the function of the cell, caused by dietary components, cancer-inducing chemicals or radiation. Normally, cells divide only when the body needs them to, and the process is under strict genetic control. When this genetic control breaks down and the cells begin dividing in an unregulated manner, a lump of excess cells forms (a tumour).

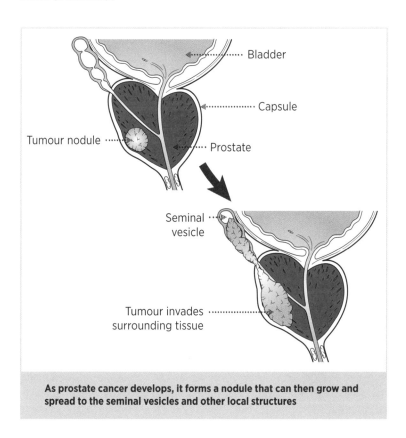

As prostate cancer develops, it forms a nodule that can then grow and spread to the seminal vesicles and other local structures

A tumour can be benign or malignant, depending on its capacity to invade healthy surrounding tissue (if it can invade, it is cancerous). Because of its capacity to invade surrounding areas, prostate cancer can spread to sites around the prostate, in which case it is said to be 'locally advanced'. It can also spread to distant sites in a process known as metastasis, which occurs as the cancer becomes more advanced.

Cancer cells can break off from the tumour in the prostate and enter the bloodstream and lymphatic system (the latter is a network of tiny vessels that drain fluid from all the organs in the body). In this way, cancer cells can spread to other parts of the body (for example, the lymph nodes or bones) and, like seeds growing in fertile soil, secondary tumours develop.

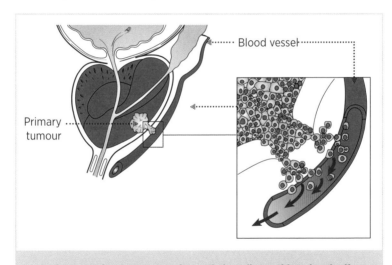

As the cancer becomes more advanced, the cells are able to break off from what is known as the primary tumour. These cells enter the blood or lymphatic system, and are transported to distant parts of the body. Once deposited at a site, the cancer cells start to grow and multiply, and new secondary cancers called metastases are formed

## Stages of cancer

The earliest stage in uncontrolled cell growth is not actual malignancy, but pre-malignancy, known medically as prostatic intraepithelial neoplasia (PIN for short). PIN is characterized by a 'heaping up' of cells within the prostate, but there is no invasion of healthy tissue at this stage. With time, however, these dividing cells may develop the ability to invade the prostate tissue. Such early signs of invasion give the pathologist examining a sample (biopsy) of prostate tissue under a microscope the clue that actual cancer has developed from the pre-malignant PIN changes. At this stage, the level of PSA (see pages 13–17) in the blood usually begins to rise – another clue that invasive prostate cancer is developing.

As cancer develops from prostate cells, when looked at under the microscope, early less aggressive cancers bear a close resemblance to normal tissue. As the cancer becomes more aggressive and potentially dangerous, these similarities are progressively lost. This process is known as 'de-differentiation' and was described by the pathologist Dr Gleason. A sample of prostate tissue is given a 'Gleason grade' according to the shape, size and structure of the cells in the sample. The grading runs from 1 to 5; the higher the number, the more dangerous the cancer (see page 26). Because the cells will not appear uniform across the tissue sample, the two most prominent regions are usually assessed, and the two grades added together to give what is known as the 'Gleason score'. Doctors can use this to estimate the risk of progression for their patients. The higher the score (from 2–10), the more potentially dangerous the cancer in terms of spread to bones and lymph nodes.

Once prostate cancer cells have developed the ability to invade tissue, they initially spread locally within the gland and then start to invade the capsule that surrounds the gland. Small tumours can be detected only by examining a biopsy of an apparently normal gland under the microscope; larger cancers can often be felt by the doctor as a firm nodule during an examination via the back passage (rectum), known as the digital rectal examination.

At first, the cancer spreads locally to tissues around the prostate, such as the seminal vesicles. Eventually, however, it can spread to more distant sites, such as the bones. The mechanisms by which cancer cells acquire the life-threatening ability to spread (metastasize) are currently the subject of intense scrutiny. Central to the process is the ability to obtain a new blood supply to provide oxygen and nutrients to the cancer cells so that they can grow (all cells have these requirements). The development of a new blood supply has been termed 'angiogenesis', and angiogenesis inhibitors, which include the infamous drug thalidomide, as well as newer agents such as Avastin (bevacizumab), provide a promising new avenue of treatment for several cancers. As yet, however, none of these has been approved for clinical use in men with prostate cancer.

## Why do some men get prostate cancer and others do not?

Overall, the lifetime risk of a man developing prostate cancer is around 12%. Your chance of getting prostate cancer depends on your personal risk factors. A risk factor is something that makes you more likely to develop a certain disease; for example, a high cholesterol level in the blood is a well known risk factor for heart disease.

The strongest risk factor for prostate cancer is increasing age. The disease rarely occurs in men under 40, but commonly affects men beyond this age. The average loss of life expectancy is about 9 years – precious retirement years for which most men have been working and eagerly anticipating all their lives.

The next most important risk factor for prostate cancer after age is family history. Like breast cancer, prostate cancer runs in certain families and has been linked to a growing number of genes (more than 70 at the time of writing). A man whose father, brother, grandfather or

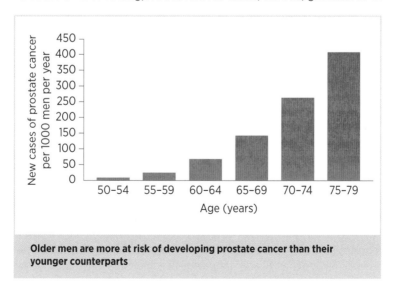

**Older men are more at risk of developing prostate cancer than their younger counterparts**

uncle has had the disease has an increased risk of developing prostate cancer compared with one without an affected relative (and the risk is higher if multiple family members have been affected). This is particularly the case if the disease developed in the close relative when he was under 60.

Race is also a factor, with men of Afro-Caribbean extraction being at highest risk. These men seem to develop a more aggressive form of the disease and at a younger age than white men. Men of Far Eastern descent seem to be relatively less likely to be affected by the disease.

## RISK FACTORS

**RISK FACTORS FOR PROSTATE CANCER:**
- Belonging to an older age group (usually 50+ years)
- Having a close family member or members who have had prostate cancer
- Having certain racial origins; for example, it is more common among Afro-Caribbean origin
- Following certain eating patterns, such as a diet high in saturated fats and red meat
- Low exposure to sunlight

## Can prostate cancer be prevented?

Clearly you cannot change your age, ancestry or race (these are 'non-modifiable risk factors'). However, several other risk factors for prostate cancer have been identified over which you can have some influence – lifestyle factors, such as diet and exercise. In addition, various dietary supplements, such as lycopenes, may offer some protection (see pages 6–8).

Geographically, prostate cancer tends to become more common as you move away from the equator: Norway and Sweden have the highest death rates from the disease worldwide. This fact points us to two further possible modifiable risk factors – low vitamin D and low exposure to sunlight, which itself helps the body to produce vitamin D. This evidence provides a good excuse for regular winter holidays in the sun! Alternatively you can take a daily vitamin D supplement.

As already mentioned, prostate cancer is characterized by an abnormal overgrowth of prostate cells. As scientists unravel the steps involved in the development of this abnormal cell overgrowth, it is possible, and indeed probable, that we will one day be able to intervene to reverse the earliest phases of the disease. A number of compounds that have this potential are currently being investigated for effectiveness and safety. One of the problems is that it is

## CASE STUDY

**KENNETH,** a moderately overweight 64-year-old university lecturer, requested a PSA check from his GP after reading about prostate cancer in the newspaper. The result came back a little raised at 5.6 ng/mL. He was referred to his local urologist who rechecked the PSA and examined his prostate, which was found to be enlarged but soft with no nodules. A biopsy was performed under antibiotic cover and local anaesthesia; it did produce some blood in the semen but no other side effects.

The biopsy revealed the presence of prostatic intraepithelial neoplasia (PIN), a condition regarded as pre-cancerous, in two out of eight cores and some inflammation in several of the remaining cores. No cancer was detected. Ken was advised to initiate lifestyle changes, with an improved diet and more exercise, and to take vitamin D and lycopene supplements and/or lycopene-rich foods, such as tomatoes, on a regular basis. He will be followed up and a rebiopsy considered if the PSA rises or if the consistency of the prostate begins to feel suspicious, or deteriorate on MRI scanning.

considerably more difficult (and expensive) to demonstrate that a given drug or vitamin is capable of preventing a disease than it is to show that it can cure a specific problem once it has developed. Because we are never sure exactly who will develop a disease such as prostate cancer, very large numbers of individuals have to be studied for many years (5, 10 or even 15) before we can be certain that a drug can safely and effectively prevent the disease from occurring.

The drug Proscar (finasteride) has been evaluated for its preventative activity. A report has revealed that 25% fewer cases of prostate cancer occurred in the men treated with Proscar at a dose of 5 mg/day. Surprisingly though, those few cancers that did occur appeared to be more aggressive in nature than those that occurred in the men not treated with Proscar. Another large study, known as REDUCE, looked at a medication that acts in a similar fashion to Proscar, namely Avodart (dutasteride). It found that the likelihood of getting prostate cancer was 23% lower in the men taking dutasteride for 4 years compared with men taking a placebo (dummy medication). Once again, though, there were concerns about a very slight increase in the number of more aggressive cancers in the group receiving dutasteride. For this reason, neither Proscar or Avodart has been licensed for the prevention of prostate cancer. Recently, reports have begun to appear suggesting that the cholesterol-lowering drugs known as statins, such as Lipitor (atorvastatin), may offer some protection against prostate cancer as well as heart disease. This is intriguing, but needs to be verified.

Other so-called chemopreventative agents will doubtless emerge as more research is undertaken. There are now more than 70 genetic mutations that appear to confer susceptability to prostate cancer. In the future we may be able to identify a sub-group of those genetically more likely to develop prostate cancer and focus preventative and early detection efforts on them.

# Treatment options for prostate cancer that has not spread beyond the gland

The most appropriate treatment for you will depend on several factors:

- How aggressive and advanced your cancer is (the grade and stage)
- Your age
- Your general health
- You and your family's own informed treatment preferences

For example, for older men with small tumours and those with other severe illnesses, often the best option is either active surveillance or 'watchful waiting'.

## Active surveillance

When you first hear of active surveillance, which entails regular monitoring but no immediate treatment, you may think 'what a cop-out', and media reports of older patients receiving second-rate healthcare may spring to mind. But active surveillance is not a second-rate option at all – it is often a way of allowing you to retain maximum quality of life, without risking shortening your life. The chance that a small slow-growing "low risk" tumour will cause problems to an older man before the end of his natural life is often relatively slight. On balance, the side effects of the other treatment options would probably cause far greater distress. As the name 'active surveillance' suggests, although you will not receive treatment, you will have regular check-ups and your urologist will monitor your condition closely with PSA measurements, MRI scans and usually repeat prostate biopsies. If the tumour shows signs of progression on PSA measurements, MRI changes or worsening results on repeat biopsy, radiotherapy or surgery may be advised.

> **TYPICAL MONITORING TIMETABLE IN ACTIVE SURVEILLANCE**
>
> **FIRST YEAR AFTER DIAGNOSIS**
> - PSA test followed by a digital rectal examination every 3 months
> - Repeat MRI scan followed by repeat prostate biopsy at 1 year
> - Case review by multidisciplinary team after 1 year's results are known*
>
> **AFTER 1 YEAR, IF NO SIGN THAT THE CANCER HAS PROGRESSED**
> - PSA test followed by a digital rectum examination every 6 months
> - Repeat MRI scan followed by prostate biopsy at 3-5 years
>
> *If your cancer seems to be growing your team will discuss active treatment options

If you choose the active surveillance option you must, for your own peace of mind, be convinced that it is right for you. Despite all the progress made in early diagnosis and treatments, a diagnosis of cancer of any kind is still distressing for the patient, and for his family and friends. It would be a rare person who, having been told that he has cancer, then manages to put the diagnosis out of his mind. It is all too easy to understand everything and feel confident that you are doing the right thing while you are in the urologist's consulting room, and then a few weeks later start to feel panicked and uneasy that nothing is being done about your condition. Remember that the whole point of active surveillance is that your quality of life remains good – if you start to worry unduly, perhaps losing sleep, then your quality of life is suffering. If this happens, pick up the phone or write to your GP or urologist and tell him how you feel. You might also find that becoming involved with a support group helps (see page 118). As well as having careful follow-up, it is important to change your lifestyle by increasing the amount of exercise you do and improving your diet (see pages 6–8).

## Watchful waiting

Watchful waiting involves monitoring for symptoms, and then treating symptoms if and when they arise. The rationale for this approach is similar to that for active surveillance, but the follow-up schedule is less intense. This approach does not offer a cure and is usually reserved for the older, less fit cancer patient.

## Radiotherapy

Radiotherapy is most appropriate for the older man whose cancer is confined to the prostate or surrounding area. But it is also suitable for the younger man whose general health precludes major surgery or in men who are worried about the side effects of surgery. With this type of treatment, radiation is applied to the affected area – the prostate and surrounding tissues – to destroy the cancer cells, leaving normal cells relatively unaffected. You may be offered one of two types of radiotherapy: external-beam radiotherapy is the most commonly used, but another method, called brachytherapy (see pages 47 – 49), is also becoming more widely available.

### External-beam radiotherapy (EBRT)

As the name suggests, a beam of radiation generated by an external source is directed at your lower abdomen. This is normally an outpatient procedure, and the most usual pattern is 20–30 minutes of treatment, 5 days a week for 6–7 weeks.

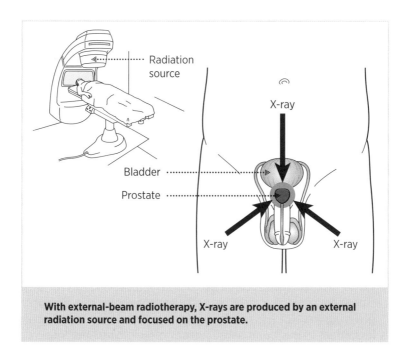

With external-beam radiotherapy, X-rays are produced by an external radiation source and focused on the prostate.

About 3 months before the radiotherapy, you will usually be given a course of hormone therapy (see pages 65-68). This shrinks the prostate tumour so that the radiation is more likely to destroy the cancer cells, which are now concentrated in a smaller area. Be aware that these hormones may affect your sex life.

A newer form of radiotherapy – conformal radiotherapy (CFRT) – has now been introduced. The use of a so-called 'multi-leaf collimator' allows more accurate targeting of the cancer and so carries a lower risk of side effects – ask your radiotherapist whether it is available in your area. The very latest form of radiation treatment is intensity-modulated radiotherapy (IMRT), which allows high doses of radiation to be precisely shaped to the individual patient's prostate. This very expensive high-technology equipment is likely to become increasingly available over the next few years.

### CyberKnife

CyberKnife uses image guidance and computer-controlled robotics to deliver with great accuracy around 1200 beams of high-energy radiation to the tumour. The aim is to target the tumour with radiation strong enough to kill the cancer cells while leaving the healthy surrounding tissue intact. In order to improve accuracy, gold seed markers are implanted prior to treatment.

The treatment duration is short, perhaps only a week, but long-term follow-up data are required before it becomes an established therapy.

### Possible side effects and risks

The main side effects of radiotherapy are bladder irritation and a need to urinate more often. Usually these effects are mild, though a very small proportion of men will be severely affected. You may also feel irritation or discomfort in and around the rectum, and notice some diarrhoea and bleeding; these effects are usually temporary, lasting only for a few weeks, but may persist for a longer time in some men. Recently, it has been reported that men who have undergone pelvic irradiation for prostate cancer have a slightly higher risk of developing rectal cancer. See your doctor if you have any bleeding from the back passage some time after treatment: a colonoscopy may be advised.

A proportion of men who have undergone radiotherapy will develop problems with erections as a result. This problem tends to develop gradually over 6–12 months, but can usually be overcome with the use of treatments such as Viagra (sildenafil), Cialis (tadalafil) or Levitra (vardenafil) or prostaglandin injections.

## Brachytherapy

Brachytherapy has become popular in the USA and is becoming available at an increasing number of centres in the UK. It involves the implantation of multiple radioactive pellets into the prostate, so the radiation is emitted from inside rather than from an external source (as is the case with external-beam radiotherapy). The pellets are left inside the patient where they gradually lose their radioactivity over the following 12 months.

Before the pellets are implanted, the radiotherapist will need to assess your prostate exactly. In order to do so, an ultrasound probe will be inserted into your rectum so that an ultrasound scan can be seen on a computer screen. The pellets – usually between 60 and 100 – are then put into your prostate using needles inserted under general anaesthetic

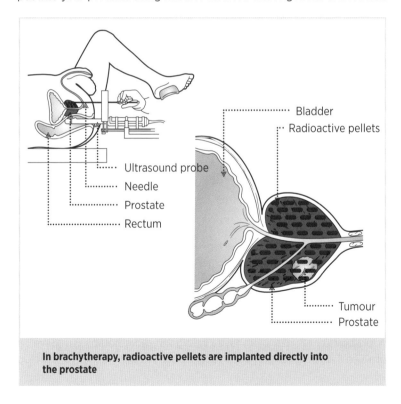

In brachytherapy, radioactive pellets are implanted directly into the prostate

through the skin between your scrotum and rectum. You will usually be fitted with a catheter to help you pass urine after the operation, which will have to stay in place for 12 hours or so, but you can normally go home within 24 hours. After brachytherapy the PSA levels gradually decline, but not usually to as low a value as after surgery.

Brachytherapy is most suitable for smaller lower-risk cancers and for small or medium-sized prostates. If a TURP (transurethral resection of the prostate; see pages 92 – 94) has been performed previously to treat BPH, the seeds cannot be sited correctly in the gland. Pre-treatment with prostate-shrinking drugs, such as LHRH analogues, can sometimes make brachytherapy suitable for men with larger glands. Brachytherapy is not appropriate for men whose cancer has spread beyond the prostate.

## Possible side effects and risks

Up to 10 years after treatment, the results appear to be good in

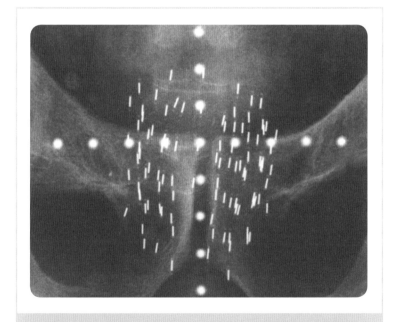

**An X-ray showing the radioactive pellets in place in the prostate. The larger dots in the shape of a cross are used to help target the prostate**

terms of keeping the PSA level down and local cancer control. As the radiation is being targeted at the prostate so accurately, urinary problems and rectal damage are possibly less common after brachytherapy than after external-beam radiotherapy. Problems with potency, though, are still common. One problem with brachytherapy is that it is difficult to treat a cancer that recurs following this treatment, as surgery in this situation is risky and more radiation cannot be given.

## Radical prostatectomy

A radical prostatectomy is a surgical procedure in which the prostate, seminal vesicles and, sometimes, a sample of some nearby lymph nodes are removed. It is quite a technically demanding operation and, as a result, is usually carried out only in certain hospitals by surgeons with particular expertise and experience. Because it is a fairly major operation, and pelvic surgery always carries certain risks, a radical prostatectomy is most suitable for otherwise healthy, younger men (generally those under 75) whose cancer appears not to have spread to the distant lymph nodes or bones.

The traditional 'open' operation is carried out under a general anaesthetic, and usually takes 2–3 hours; you should expect to stay in hospital for 3–7 days. An 8–10 cm lateral or vertical cut will be made through your abdomen above the pubic bone (or less commonly through the perineum), and your prostate and seminal vesicles will be removed. Samples from the lymph nodes nearest to your prostate may also be taken to check whether the cancer has spread. The so-called cavernous nerves, which lie close to the prostate and are important for achieving an erection, will be identified and the surgeon will take particular care not to disturb them (this may not be possible if the cancer has spread very close to the nerves); this is called a nerve-sparing approach. A catheter will be inserted into the penis so that urine can drain while the join (technically called the anastomosis) between the bladder and urethra heals. The catheter will usually have to stay in place for up to a fortnight (so you will often have to keep it for a week or so after you go home). The scar from the operation heals quite quickly and after a few months will be almost invisible.

You will need to take it easy when you return home from hospital: the usual period of convalescence is 6–8 weeks, but you may still feel tired even after this time. Avoid lifting heavy objects for several months.

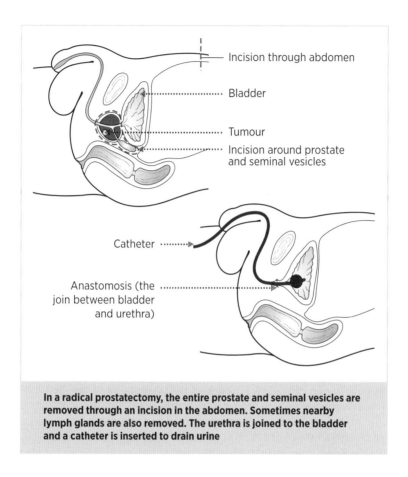

In a radical prostatectomy, the entire prostate and seminal vesicles are removed through an incision in the abdomen. Sometimes nearby lymph glands are also removed. The urethra is joined to the bladder and a catheter is inserted to drain urine

Some guidelines as to what you should and should not do after the operation are shown in the table opposite.

## Laparoscopic and robotic radical prostatectomy

Recent technological developments have enabled the prostate to be removed using a tiny telescope and 4–6 small incisions ('minimally invasive' or 'keyhole' surgery). The abdominal wall is punctured and the abdominal cavity is distended with gas (carbon dioxide). The surgery is then performed by a surgeon who is guided by the magnified image on a television monitor. The advantages of this technique include reduced blood loss and a quicker recovery time, but the disadvantages may be a somewhat longer operating time and the difficulty in training surgeons to perform what is a technically demanding procedure.

The latest development is the use of the da Vinci® robot to assist with the laparoscopic operation. This device allows three-dimensional visualization at 10 times magnification and very precise control of movement, which does seem, in experienced hands, to reduce blood loss and perhaps enable better preservation of the nerve bundles that are important for erections.

## AFTER A RADICAL PROSTATECTOMY

### RETURNING TO WORK
- Possible after 4-6 weeks
- A longer period of absence may be necessary if your job involves heavy lifting
- Your doctor will give you a sick note

### DRIVING
- Do not drive for 2–4 weeks after the operation

### SEXUAL ACTIVITY
- You can attempt to have sex whenever you feel ready
- Orgasm can usually be reached, but there will be no ejaculate and your erection will be weak initially. A low daily dose of Viagra or Cialis may help your sexual recovery, moving to higher doses as and when required. Prostaglandin injections may also be helpful

### DRINKING
- Try to drink more (non-alcoholic drinks) than you would do normally. The resulting increase in the volume of urine produced can help protect against infection
- You can drink alcohol (but this may affect your continence so should be in moderation)

### EXERCISE
- Rest as much as possible for the first 2 weeks
- Avoid any heavy work, such as lifting, carrying or digging, for several months
- Sports and exercise can be resumed after about 1 month, but be guided by how you feel and start off very gently (swimming is a good exercise to begin with)

The American surgeons who originally developed the technique recently reported that more than 80% of their patients were able to have satisfactory intercourse some months after surgery – but this is optimistic! The number of robots in action in the UK is likely to increase, as results are impressive and surgeons are very enthusiastic about this new technology.

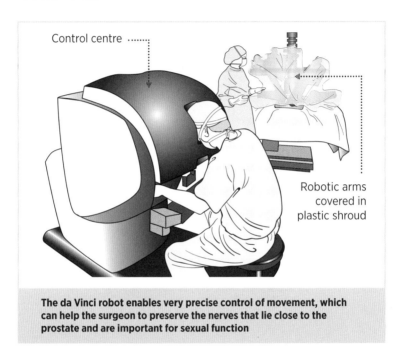

The da Vinci robot enables very precise control of movement, which can help the surgeon to preserve the nerves that lie close to the prostate and are important for sexual function

## PSA level after surgery

After the operation, your PSA level will be checked every 3 months for at least a year. It should drop below 0.1 ng/mL, but this will depend to some extent on the laboratory that performs the analysis; some laboratories have machines that only measure PSA as low as 0.2 ng/mL, whereas others have machines that can measure as low as 0.1 or even 0.01 ng/mL.

If your PSA starts to rise because the cancer has not been completely removed (remember that almost all prostate cancer cells secrete PSA), you will usually need further treatment. Slight flickers in the PSA may occur, however, and do not always need treatment.

## When further treatment is needed

One-tenth to one-third of all men who undergo radical prostatectomy are found to have cancer that has spread to or beyond the margin of the prostate. This finding, which is known as a 'positive margin', is particularly likely in men whose PSA level is above 10 ng/mL. As a consequence, the operation will sometimes not be 100% successful in these men as the cancer has not been wholly removed from the body. If this is the case for you, your doctor may recommend a 'mop-up' course of radiotherapy or some long-term drug therapy with anti-androgens and/or LHRH analogues (see pages 66 and 68).

## Possible side effects and risks

A radical prostatectomy, even using the latest laparoscopic and robotic technology, is major surgery and, as such, has side effects that you should consider carefully when deciding whether this is the appropriate course of action for you. For men who may have wanted children, infertility from the surgery needs to be talked through thoroughly with their doctor and partner. Sperm banking is one option that could be considered.

Many men also experience a degree of temporary stress urinary incontinence after the operation. For most, incontinence is mild – a leakage of a small amount of urine on coughing, for example. A very small proportion of men have severe incontinence requiring further treatment, but very few have a permanent problem, other than having to wear a small pad for security.

Impotence or erectile dysfunction (difficulty achieving an erection) is another side effect and affects many men who have undergone a radical prostatectomy. The risk is reduced where a surgeon uses a nerve-sparing approach but, even so, potency cannot be guaranteed. Although impotence can usually be treated reasonably effectively, the surgeon should discuss this with you in detail before surgery, and you should discuss it with your partner. Recent evidence suggests that early active rehabilitation using Viagra (sildenafil) or similar agents, such as Cialis (tadalafil) or Levitra (vardenafil), can help to restore sexual function after surgery. Prostaglandin injections at the time of desired intercourse are almost always effective in this situation.

Internal scarring from the operation is a further potential complication.

If your urine flow deteriorates after surgery, it may mean that you will have to undergo dilatation (stretching) of the join between the bladder and urethra; this is usually curative, but sometimes has to be repeated. Some patients will require a period of self-catheterization to ensure that the join between the bladder neck and the urethra remains wide open as it heals. The risk of this is now thankfully much lower after laparoscopic or robot-assisted surgery. On the positive side, for men who have BPH as well as prostate cancer, radical prostatectomy can potentially offer a 'double cure' as the prostate, the source of the BPH symptoms of frequency and poor flow, is removed.

The risks associated with radical prostatectomy are those generally associated with any major surgery – blood loss or blood clots, an adverse reaction to the general anaesthetic, and infection. With both laparoscopic and robotic, bowel injury may occur, but is very rare.

Your chance of experiencing side effects and the likely success of the operation are governed largely by the expertise of your surgeon. If you are offered this operation, you should ask your urologist a number of questions.

- How many radical prostatectomies have you performed (more than 100 is a respectable number) and how many in the last year?
- What were the results of these operations, in terms of removing the cancer, and what was the proportion of patients who were free from the major side effects of impotence and incontinence at, for example, 1 year after surgery? (More than 50% of patients younger than 60 years of age able to have intercourse, with or without medication, and fewer than 2% of patients with severe incontinence are good results.)
- Will you be performing my surgery personally?
- Will it be open, laparoscopic or robotic?
- Will you be there to help if I have post-operative problems?
- When will the pathology report be available and will you be there to explain it to me?
- How often will I be seen for follow-up?
- Will I be given help to deal with sexual rehabilitation and possible stress incontinence?

If you are not happy with the answers, you can always seek a second opinion.

## Surgery or radiotherapy?

Ultimately, when you have weighed up the pros and cons with your urologist, the choice will be yours and that of your immediate family. The risks associated with radical prostatectomy or radiotherapy and a summary of the pros and cons of each is shown in the table overleaf.

### How do I make the choice?

Until the results of ongoing studies are available, it will not be known for certain which is the safest and most effective treatment for localized prostate cancer. Until then, there will be a choice of treatments, which the patient must decide for himself. Critical to this choice is the confidence that you feel in your doctor and his team, so it is important to find an experienced specialist team and weigh up the pros and cons with them, before deciding for yourself which treatment is right for you.

## The long-term picture

Long-term studies provide information on the prospects of men who have undergone these procedures. While many men want this kind of information, it is important not to take the figures given here too much to heart without discussing your own individual circumstances with your urologist. Progress in medicine means that patients' long-term prospects are improving all the time, and in due course the results of ongoing clinical studies will resolve many controversies.

### Active surveillance

The likelihood that your cancer will spread depends, as has already been said, on the nature of your cancer (that is, how aggressive it is). For men whose cancer has a low Gleason score (i.e. well differentiated tumours), the 10-year survival rate is 87%, which means that, after 10 years, 87 men in 100 will not have died from prostate cancer. With more aggressive cancers (those with higher Gleason scores), the survival rate drops considerably (the 10-year survival rate for men with poorly differentiated tumours has been put at 26%). Active surveillance is often a good option to start with as more active treatment can always be instituted if signs of cancer progression develop. However, careful follow-up and regular testing are essential.

## THE PROS AND CONS OF RADICAL PROSTATECTOMY VERSUS RADIOTHERAPY

### Radical prostatectomy

**Pros**
- Offers a cure for tumours confined to the prostate
- Can now be carried out by keyhole surgery with robotic assistance
- Allows the doctor to stage your tumour accurately
- Coexisting BPH is treated
- Your PSA level should become undetectably low
- You are likely to feel reassured about your condition after the operation
- Monitoring for cancer reappearance is easy
- Radiotherapy can be given afterwards if the cancer returns

**Cons**
- Major surgery
- Small risk of severe bleeding associated with operation
- Success/side effects depend on the skill of the urologist
- Possible side effects (see text)

### Radiotherapy/ brachytherapy

**Pros**
- Offers a potential cure
- Avoids prolonged catheterization and surgery
- Given on an outpatient or short-stay basis
- Hormone therapy can increase the chance of success

**Cons**
- Treatment is prolonged (6 weeks external-beam radiotherapy)
- It is relatively difficult to assess whether the treatment has been successful
- Accurate staging is not possible
- Coexisting BPH is untreated
- You may feel more concerned about the possible chance of success afterwards
- Your PSA level may not drop to very low levels
- Repeat radiation treatment is not possible
- Surgery after radiotherapy carries greater risks and is only suitable for a very few selected cases
- Possible side effects (see text)

## Outcomes from radical prostatectomy

More than 80% of men who have this operation are alive 10 years afterwards, and 60% are still alive at 15 years. An important Scandinavian study compared the long-term outcomes of men who chose watchful waiting with those treated by radical prostatectomy. The results suggest that radical prostatectomy is the treatment option most likely to offer a complete cure for younger men with higher-risk tumours, as it physically removes both the cancer and the entire prostate from the body, making recurrence and spread to the bones much less likely. Another study published recently also showed a definite eight-year survival advantage in patients with higher-risk cancers treated with surgery.

Results from a study by Pound and colleagues confirm that 82% of men undergoing radical prostatectomy at Johns Hopkins Hospital in Baltimore (USA) were free of recurrence at 15 years (as determined by PSA measurement). The study also offers some comfort to those men whose PSA level rises years after the operation. As we have

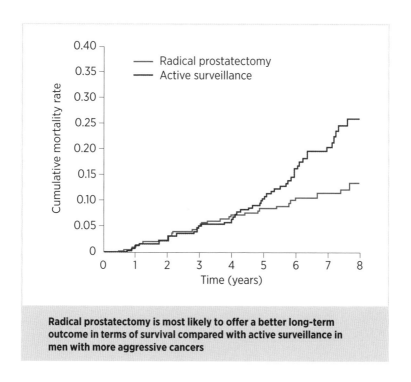

**Radical prostatectomy is most likely to offer a better long-term outcome in terms of survival compared with active surveillance in men with more aggressive cancers**

already said, after a radical prostatectomy, your PSA level drops to an undetectable level and, if it starts rising again, it can signal cancer recurrence. The data from the study in question indicate that, though this is the case, the cancer spreads in only around one-third of men with an elevated PSA. Furthermore, unless a man had a particularly aggressive cancer (in which case his PSA level would tend to rise relatively quickly after the operation), the spreading cancer would not become life-threatening for several years, and would likely be amenable to treatment with radiotherapy or LHRH analogues (see pages 65 – 68).

## Outcomes from radiotherapy

At best, the survival rates with radiotherapy are comparable with those associated with radical prostatectomy. Several published studies have put the 15-year survival rates at 40–60% (that is, in a group of 100 men, between 40 and 60 will still be alive after 15 years). Recent data suggest that LHRH injections or anti-androgens to shrink the prostate ahead of radiotherapy can significantly increase the chances of successful treatment (see page 68).

The risk of serious side effects with radiotherapy is decreasing as improved technology means that the cancer-destroying rays can be targeted more accurately at the cancer, leaving adjacent structures, such as the rectum, undamaged. Proponents of brachytherapy also report improving results as techniques and patient selection are enhanced. However, problems with potency are still frequently encountered after radiotherapy and, in fact, are much more common when this treatment is combined with hormone therapy.

A progressive rise in PSA after either external-beam radiotherapy or brachytherapy does suggest that recurrence has occurred. Although salvage surgery is technically feasible in some cases, it is often difficult and associated with a high complication rate. Sometimes, cryosurgery or high-intensity focused ultrasound (HIFU) can be used to destroy the residual cancer (see the following section), but more commonly hormone therapy is used.

# Experimental options

Newer treatments such as cryotherapy and HIFU may become useful but are currently available only as part of a clinical trial.

## Cryotherapy

Cryotherapy uses freezing to destroy the prostatic tissue. An ultrasound probe in the rectum enables the position of the prostate to be seen on a computer screen. A number of 'cryogenic' probes are then inserted into the prostate, and liquid nitrogen is circulated to reduce the temperature to around −180°C. At this temperature, the tissue surrounding the probes is destroyed. The urethra is protected by circulating warm water through a catheter. Some studies have reported survival rates similar to those achieved with radical prostatectomy, but others have described rectal and urethral damage, which can be difficult to repair. No long-term randomized controlled trials to compare cryotherapy with established treatments have yet been carried out. Currently, it is mainly used as a treatment for prostate cancer that has recurred after radiotherapy, since other treatment options in that situation are limited and the technique does offer potential cure. Some surgeons are, however, starting to use it as a primary treatment for patients with locally advanced cancers who wish to avoid radiotherapy. Cryotherapy is technically demanding so it is important that the team treating you has wide experience of this procedure.

## High-intensity focused ultrasound (HIFU)

HIFU is a technology that allows ultrasound waves to be focused on prostate cancer cells. It involves the insertion of an ultrasound probe into the rectum under anaesthesia and then the destruction of the cancer cells by ultrasound energy; the treatment can take up to 3 hours. It can be used to treat both newly diagnosed cancers and recurrences after radiotherapy. Currently it is possible to have this treatment only as part of a clinical trial; it is not yet accepted by the National Institute for Health and Clinical Excellence (NICE) for general use in the NHS. Initial results look reasonably encouraging, since the PSA levels seem to decline and side effects are not prominent, although a catheter is required for several days and sometimes longer because the prostate swells in response to therapy.

Damage to the bladder and rectum have been described as a result of HIFU, as well as incontinence, so you should ensure your surgical team has extensive experience with this technique if you are offered this as an option. Much longer-term follow-up and trials comparing it with surgery and radiotherapy will be required before HIFU can be regarded as a mainstream treatment. 'Focal' treatment, when only half the prostate is treated by ultrasound, is possible – and may help to preserve potency – but again, longer-term follow-up is necessary before this is accepted as a standard therapy.

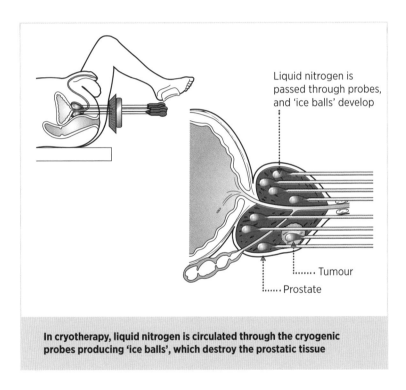

In cryotherapy, liquid nitrogen is circulated through the cryogenic probes producing 'ice balls', which destroy the prostatic tissue

# CASE STUDY

**DOUGLAS,** a 58-year-old banker with an uncle and a father who had both suffered from prostate cancer, went to his GP complaining of the need to get out of bed several times at night to pass urine. A PSA test was requested which came back elevated at 7.8 ng/mL. He was referred to his local urology department and underwent a biopsy of the prostate, the results of which revealed Gleason grade 4 + 3 = 7 prostate cancer on the left side of the gland. An MRI scan showed no evidence of disease outside the prostate.

He researched his treatment options on the internet and discussed his case with various helplines. He was attracted to both high-intensity focused ultrasound (HIFU) and brachytherapy as he was nervous about surgery. Further investigations confirmed that his prostate was enlarged and that he was not emptying his bladder properly, and so he followed the advice of his urologist and underwent laparoscopic prostatectomy.

He made a rapid recovery and, on removal of the catheter, was able to pass urine with an improved flow and was completely continent. Examination of the tissue removed confirmed a 2 cm tumour on the left that had been completely removed. Currently, his PSA is undetectable and he is able to function sexually with the help of sildenafil (Viagra).

# If prostate cancer has spread or recurs after treatment

## Locally advanced disease

If your cancer has spread outside your prostate, but has not yet spread to the lymph nodes close by or to more distant locations, such as the bones, it is described as being 'locally advanced'. (In the TNM staging system, this state is known as T3-N0-M0.) The treatment options for such disease are:
- Active surveillance or watchful waiting (for older, less fit men, as before)
- Hormone therapy
- Intermittent hormone therapy
- Hormone therapy followed by radical prostatectomy
- Hormone therapy followed by radiotherapy
- Anti-androgen alone (monotherapy).

### Active surveillance

The rationale behind adopting the approach of active surveillance has been outlined earlier (see pages 43 – 44). However, it is important to realize that at this stage, because the cancer is more advanced, it is likely to cause symptoms and become life-threatening more quickly than a low-grade cancer that is still confined to the prostate. Active surveillance for locally advanced prostate cancer is therefore mainly applicable to older, less fit men with a shorter life expectancy.

### Hormone therapy

Hormone therapy is sometimes called 'cytoreduction', and has been touched on in the previous section. There are several ways by which this can be achieved:

- LHRH analogues, which initially stimulate then block production of testosterone; LHRH is a naturally occurring hormone, and the 'analogue' part of the name means that this is a synthetic form with a structure similar to the natural form
- LHRH antagonists, which block the production of testosterone

- anti-androgens, which block the action of the male hormone testosterone in the body

Testosterone, an androgen or male hormone, is produced in the testicles and has the effect of stimulating cancer growth. The aim of hormone therapy is to reduce the effect of testosterone by switching off testosterone production (the LHRH analogues or LHRH antagonist) and/or by dampening its effects on the cancer (the anti-androgens). The overall effect is that the tumour size is reduced and the progression of the tumour is delayed (hormone therapy does not offer a complete cure, however).

Usually, implants containing an LHRH analogue are inserted by injection at monthly, 3-monthly or even longer intervals. Your body may react to the first injection by initially increasing the amount of testosterone it makes – this is the so-called 'flare' effect. To counter this, you will probably be given an anti-androgen, such as bicalutamide, to take a few days before and then continue for several weeks at the beginning of treatment with the LHRH analogue. The flare phenomenon does not occur with an LHRH antagonist such as degarelix.

### Possible side effects
As a consequence of stopping the production of testosterone, men receiving an LHRH analogue lose their sex drive and are unable to achieve an erection. This is gradually reversed if the drug is stopped. Some men also experience hot flushes – these may be eased by low doses (50 mg/day) of Cyprostat (cyproterone acetate). Although anxieties about osteoporosis, cardiac problems and effects on the brain have been raised, these are seldom encountered as a clinical problem.

Anti-androgens may cause mild stomach upsets and diarrhoea. Rarely, they can have a harmful effect on your liver (so you will need regular blood tests while you are taking these tablets).

### How effective is hormone therapy?
Hormone therapy alone reduces the tumour size and slows the cancer progression in around 80% of men with locally advanced disease. It does not destroy all the cancer cells, so the cancer is not cured, but its progression is significantly delayed and the effects of other treatments, such as radiotherapy, are enhanced.

### Intermittent hormone therapy

Intermittent hormone therapy is a newer approach to hormone therapy. An LHRH analogue is given for about 36 weeks and then discontinued (providing the PSA level has dropped down to a normal value). When the PSA level returns to a predetermined level, the hormone treatment is started again. Some doctors believe that this might make the cancer cells susceptible to the drug for longer than they would be if treatment was continued without a break. Studies looking at the long-term safety and effectiveness of this approach are mainly positive, but for the moment it is still experimental.

## CASE STUDY

**ANDREW,** a 78-year-old retired decorator, went to see his GP complaining of difficulty passing urine. His PSA was found to be raised at 11.7 ng/mL and on examination his prostate felt hard and irregular. He was referred to his local urology department, and prostate biopsy as well as bone and MRI scans were arranged. These confirmed a Gleason score of 4 + 4 = 8 prostate cancer with local spread to the seminal vesicles, but no involvement of either lymph nodes or bones.

Treatment options were discussed and surgery was discounted because of his age and the stage of the cancer. He opted for a course of conformal external-beam radiotherapy preceded by 3 months' hormone treatment with an LHRH analogue to reduce the size of the tumour. He tolerated the radiation treatment well, although he did complain of loss of libido and some rectal bleeding towards the end of therapy.

For the 2 years since then his PSA has remained stable and suppressed, and he is being followed up in clinic.

### Hormone therapy followed by radical prostatectomy
Some doctors believe that shrinking the tumour with hormone therapy before carrying out a radical prostatectomy increases the chance of removing all the cancer. This approach has been tested in long-term studies. The latest data suggest, however, that there are no concrete long-term advantages to having hormone treatment before surgery, so this approach is not generally recommended.

### Hormone therapy followed by radiotherapy
Again, studies are being carried out to see whether hormone treatment before radiotherapy gives better results than radiotherapy alone. In this case, the results are encouraging, suggesting that the hormone treatment does indeed offer a benefit in terms of curing, or at least delaying its progress. This is probably because the shrunken tumour is more susceptible to the anti-cancer effects of ionizing radiation. In men at higher risk of recurrence, the hormone therapy is often continued for several years after the initial treatment.

### Anti-androgen monotherapy
There is now scientific evidence that an anti-androgen drug alone (i.e. monotherapy) can also help to slow the progress of advanced cancer, particularly when bone metastases are not present. The advantage of this approach is that anti-androgens have less effect on sex drive and are less likely to cause erectile dysfunction than the long-acting injectable LHRH analogues. Breast tenderness and enlargement can occur but, although these side effects can be troublesome, they can usually be prevented by a short course of radiotherapy to the nipple areas. Liver function is only rarely disturbed by agents such as bicalutamide, but blood testing should be performed to be sure. Worries in Scandinavia about cardiac side effects have not been borne out by studies in other countries, so this treatment is considered by doctors to be completely safe.

## Metastatic disease
Once prostate cancer has spread secondarily to the lymph nodes and to distant sites, most frequently the bones, it is referred to as metastatic disease (the metastases are the secondary growths that occur at the distant site); in the TNM staging system, this state is known as T3–N1–M1. This is an advanced form of cancer, and one that is associated with a relatively poor outlook, but there is no need to give up hope, especially nowadays.

This stage of cancer can still be treated and progression of the disease can be delayed for several years. The treatment options are:

- orchidectomy (surgical removal of both testicles)
- hormone therapy with LHRH analogues
- 'maximal androgen blockade', which is hormone therapy with a combination of LHRH analogue and anti-androgen.

## Orchidectomy

Orchidectomy is a surgical procedure in which both the testicles are removed. The reasoning behind this is that, as testosterone is produced in the testicles, their removal stops its production altogether. Most men (more than 80%) respond positively to this treatment, with the progression of their cancer slowing markedly for around 18 months and sometimes much longer.

The operation is straightforward and is performed under a local or general anaesthetic in around 30 minutes. The scrotal sac is opened and the testicles are snipped out. In selected patients, silicone testicular prostheses (implants) may be inserted to improve the cosmetic result. You may be allowed out of hospital on the same day, although often your surgeon will want you to stay in overnight to check for bruising. You must take things easy for a week or two, and you should also take regular baths or showers to keep the wound clean. Afterwards, the scrotum will look a little bruised, and later somewhat shrivelled and empty, unless testicular prostheses have been used.

Although the operation seems rather drastic, and some men are concerned about 'castration' and the appearance of their scrotum afterwards, it is a one-off procedure and so avoids the need to take a prolonged course of hormone therapy. However, it is nowadays a rather uncommon way to treat prostate cancer as medical therapy tends to be preferred.

### Possible side effects and risks

As your body will be unable to produce testosterone after the operation, you will lose your sex drive and be unable to achieve an erection. You will also be infertile. These effects are irreversible, so consider the implications very carefully before consenting to an orchidectomy. Potential complications of the surgery are relatively few, but bruising, blood clots and infections may occur in some men.

Hot flushes may result from the hormone changes in your body. You will not become 'feminized' or find that your voice changes, but you may notice that you lose some body hair and may have to shave rather less often. There is also often a change in skin texture and a theoretical risk of the brittle bone disorder known as osteoporosis.

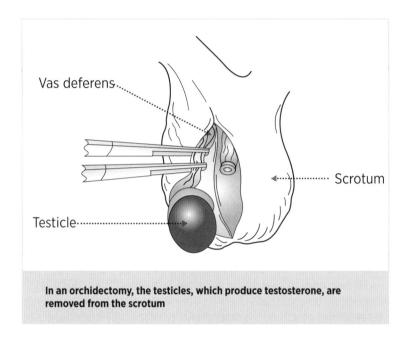

In an orchidectomy, the testicles, which produce testosterone, are removed from the scrotum

## Hormone therapy

LHRH analogues (see pages 65 – 66) achieve the same result as removal of the testicles by blocking the production of the male hormone testosterone, and thus reducing the stimulation of cancer growth. LHRH analogues, such as Zoladex (goserelin), are usually administered as an implant, which is injected just under the skin of your abdomen. The procedure is repeated every month or 3 months.

As with orchidectomy, a high proportion of men (more than 80%) respond to this treatment and the beneficial effects usually last for around 18–36 months. In terms of effectiveness and safety, there is little to choose between hormone therapy and orchidectomy, but most men prefer the former.

### Possible side effects and risks
At first, the LHRH analogue actually increases testosterone production for a few days. Bone pain may increase as a consequence, and urinary symptoms may worsen. This is known as the 'flare' phenomenon. There is even a remote risk of the cancer causing pressure on the spinal cord and thus paralysis. To counter these effects, anti-androgens such as bicalutamide are usually given for 2 weeks before and then for the first 2–6 weeks of LHRH analogue treatment; these effectively block the effect of testosterone on the cancer. Use of the LHRH antagonist degarelix avoids the risk of flare and the need to use anti-androgens when starting treatment.

### Maximal androgen blockade
Maximal androgen blockade combines the use of LHRH analogues with long-term anti-androgens. Whether or not this approach is superior to that using LHRH analogues only or orchidectomy is not entirely clear. Some studies show men respond for a longer length of time with this treatment, while others have failed to show such an effect. Many doctors do have confidence in this approach, though, and feel that it is particularly appropriate for younger, relatively fit men with advanced prostate cancer.

### Possible side effects and risks
As outlined previously, treatment with LHRH analogues results in a loss of sex drive and impotence. Hot flushes can also be a problem but sometimes respond to treatment with Cyprostat (50 mg/day). The other part of the treatment, anti-androgens, may upset your stomach and can sometimes cause diarrhoea.

## Bone-related problems
### Bisphosphonates
Since prostate cancer frequently spreads to the bone, a class of drugs known as bisphosphonates, which act to stabilize the skeleton and reduce bone loss, may be helpful. A study has demonstrated that Zometa (zoledronic acid) administered by intravenous infusion every 3 weeks can delay the development of skeletal problems, such as fracture, by up to 5 months. Side effects of this treatment are relatively minor; some patients develop a flu-like illness during the infusion but this is usually short-lived. Rarely, a problem with the jawbone can develop. More and more men with advanced prostate cancer are now

being offered this treatment option, and studies are under way to determine whether bisphosphonates may even prevent metastases in the bone developing in the first place.

### Denosumab

Denosumab is a therapeutic antibody that targets a protein involved in bone 'remodelling', a naturally occurring process in which old bone is removed and new bone is added. By targetting the protein, denosumab reduces the removal of old bone and helps protect against bone-related problems in metastatic prostate cancer. Recent evidence suggests that denosumab may be superior to Zometa in preventing skeletal complications.

## Recurrence

Almost inevitably, cancers that initially respond to the hormonal treatments eventually begin to grow again (the diagram on the next page explains why this happens). This stage of prostate cancer is often referred to as hormone-relapsed prostate cancer. The PSA value starts to rise again.

If you reach this stage, your doctor may recommend one of the following treatment options:

- modifying existing hormonal therapy by adding or withdrawing anti-androgen
- cytotoxic chemotherapy (drugs that destroy the cancer cells)
- hormone therapy (this is different from that discussed earlier)
- another form of treatment that aims to prevent stimulation of further growth of the cancer.

### Cytotoxic chemotherapy

Cytotoxic chemotherapy is an option, but the drugs used can have side effects, such as sickness and hair loss. Improvements to therapy mean that these are less frequent these days. Increasingly effective chemotherapy drugs are now available, so if your doctor discusses this with you, ask what side effects you might expect and whether it is possible to counter them effectively. Oncologists rather than urologists are experts in this area.

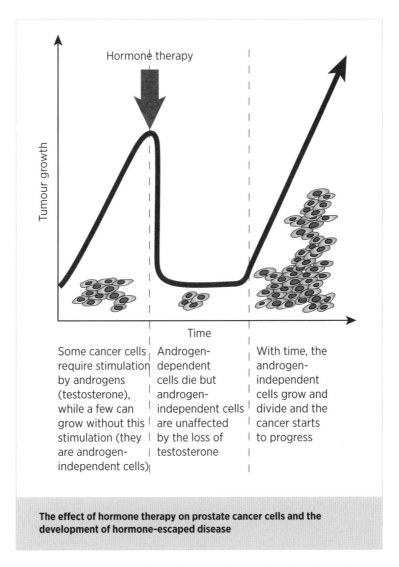

**The effect of hormone therapy on prostate cancer cells and the development of hormone-escaped disease**

So what is the point of these drugs? It is possible that chemotherapy might give you an extra few months or even years, and if the side effects are minimal or can be overcome, you might feel that this option is worthwhile, particularly as they have also been shown to improve symptoms and quality of life. Medications given intravenously in 3-weekly cycles, such as Taxotere (docetaxel), have been shown to improve survival rates. Taxotere given every 3 weeks may result in

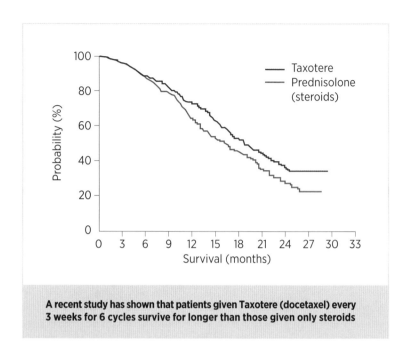

**A recent study has shown that patients given Taxotere (docetaxel) every 3 weeks for 6 cycles survive for longer than those given only steroids**

a sharp reduction in PSA values as well as an improvement in quality of life. Side effects include nausea, vomiting, hair loss and a reduction in the white cell count in the blood, known as leucopenia – your oncologist will discuss the latest treatments with you and organize treatment in an oncology centre. A second-line chemotherapy agent, cabazitaxel, has been approved for use in the USA and Europe and can be used when Taxotere begins to lose effectiveness.

## Hormone treatment

Oestrogens (female hormones) may offer some benefit at this stage of your disease. They appear to be able to reduce stimulation of cancer growth and they may also damage the cancer cells directly. The reason that oestrogens are not used in earlier disease is that they can have some potentially serious side effects, such as nausea, blood clots and other cardiovascular complications, such as heart attacks or even strokes. Many doctors advise that you take a low dose of aspirin (75 mg) if you take oestrogen-based drugs, in order to help overcome the potential cardiovascular side effects. Oestrogens should certainly not be used if you have previously had problems with deep vein thrombosis, pulmonary embolism or heart failure.

## Other treatments

There are a number of so-called 'growth factors' in the body that stimulate the progression of prostate cancer. Blocking the action of these growth factors should potentially block their stimulatory effects on the cancer. However, the drugs that are being developed with this aim are very new and are still under investigation. Angiogenesis inhibitors such as Avastin (bevacizumab) have already been mentioned (see page 38); immunotherapy with Provenge (sipuleucel-T) holds considerable promise (see pages 80 – 81), although it is very expensive and not available in the UK at the time of writing. Again though, if you do eventually reach this stage, knowledge of the effectiveness and safety of these drugs may then be such that your doctor is able to prescribe them for you.

## Abiraterone

A new drug, abiraterone, has been developed at the Royal Marsden Hospital, London. It is a hormone therapy that blocks an enzyme involved in the production of testosterone. It is now approved for use in the UK and early results are very encouraging. It can be given orally but it does have side effects and has to be combined with a steroid such as prednisolone. Current studies have involved men with very advanced disease. The role of abiraterone in the treatment of earlier stage prostate cancer is not clear, but more studies are planned.

## Enzalutamide

Another new drug, enzalutamide, has been specifically developed for the treatment of prostate cancer that has relapsed after hormone treatment. It works by blocking the androgen receptors within cancer cells that stimulate tumour growth. It is now licensed for men with hormone relapsed prostate cancer and can be used pre and post-chemotherapy.

## Palliative care

Palliative care aims to provide you with support to make you feel free from pain, comfortable and composed in the final stages of the illness. Over recent years, considerable progress has been made in this area, and medical opinion now holds that no patient need feel the pain or discomfort that was characteristic of the last stages of this cancer in bygone years.

If your cancer progresses to this stage, you will usually be assigned a palliative care team – specialist doctors and nurses who have considerable expertise and experience in this area, and who will support you and your family. You will have opportunities to talk to members of the team about your care, and you should discuss any medical, social or financial worries that you have. Macmillan and Marie Curie nurses can be very helpful and will visit you at home.

Patients with very advanced prostate cancer tend to experience bone pain, and you may be given strong and effective painkillers to help overcome this. In addition, you might have radiotherapy (either as a short course or a one-off). Another effective method of alleviating bone pain is with injections of a radioactive substance known as strontium, but more often alpharadin is used these days.

If you are offered radiotherapy, make sure that you know whether it is likely to result in other side effects, such as nausea and tiredness, so that you can weigh up the advantages and disadvantages in the light of all the facts and your own circumstances. In this situation, the first consideration of the medical team should be to preserve your dignity and help your family and friends to support you, ideally in your own home. The palliative care teams are geared up to do just this.

The charity Prostate Cancer UK has a "Survivorship" programme designed to help prostate cancer patients and their relatives and carers deal with the various problems, anxieties and perturbations associated with prostate cancer and its treatment.

## CASE STUDY

**BILL,** a 65-year-old retired driver who hails originally from Jamaica, went to see his GP complaining of tiredness, weight loss and low back pain. X-rays of the spine suggested the presence of metastatic cancer and his PSA result came back at 556 ng/mL.

He was referred urgently for prostate biopsy, which revealed a Gleason score of 5 + 4 = 9; a bone scan confirmed secondary spread particularly involving the spine. Bill was informed of the diagnosis in the presence of his supportive family and agreed to commence androgen ablation (i.e. hormonal) therapy. After 5 days of anti-androgen therapy, a 3-month depot injection of an LHRH analogue was given which quickly relieved his back pain and improved his general health. His PSA also fell quickly to a lowest value of 3.8 ng/mL, but then started to rise. At this point, Bill was referred to his local oncology department, and further blood tests and scans were arranged that confirmed that hormone relapse had occurred. He was treated with 3-weekly infusions of Taxotere (docetaxol) and Zometa (zoledronic acid) with some improvement in his well-being and a substantial fall in his PSA. He remains under follow-up and supported by his family and the palliative care team, including a Macmillan nurse at home. If a further rise in PSA occurs, treatment with cabazitaxel, enzalutamide or abiraterone may be considered. The new agent alpharadin that targets bone metastases in prostate cancer patients is another option in this situation.

# Prostate cancer: the future

The prospects for significant progress in prostate cancer in the near future are now better than ever. We can look forward to more effective prevention, earlier diagnosis, better staging, and more effective and less toxic therapy. A number of current research endeavours to improve our understanding of the disease may well translate into improved quality of life and improved survival prospects for those affected by prostate cancer.

## Chemoprevention

In the future, it may be possible to prevent prostate cancer. The benefits of the 5-alpha-reductase inhibitor Avodart (dutasteride) have been assessed in the large REDUCE trial, which showed a 23% reduction in prostate cancer diagnosed on biopsy. Unfortunately, however, a small number of patients taking this medication developed high-risk cancer. Several other agents, such as statins which are currently used to lower cholesterol, appear to show promise, but require further research to ensure their safety and effectiveness.

## Genetics and targeted screening

The recent discovery of more than 70 so-called 'prostate cancer susceptibility genes' raises the possibility of targeted screening of those men who have the highest risk of developing the aggressive form of prostate cancer. This is likely to involve close surveillance by PSA testing, MRI scanning and prostate biopsy.

## Better diagnosis

Earlier detection, while the disease is still curable, is already a reality as a result of PSA testing. In the future, new tests or variations of existing tests will continue to improve the ability of doctors and surgeons to distinguish early prostate cancer from BPH. Recently, a new test for prostate cancer called the PCA3 (prostate cancer antigen 3) test has been described, and several others are in the pipeline.

### PCA3 test

Unfortunately, the PSA test is not reliably specific for prostate cancer, so some men who have a biopsy based on their PSA result are found not to have cancer. A gene called PCA3 ('prostate cancer antigen 3') has been highlighted in the search for another, more specific, marker of prostate cancer, and early work shows that finding a man's 'PCA3 score' may be helpful. Prostate cancer cells have higher numbers (or copies) of the PCA3 gene and, because prostate cells are shed into the urine, the levels can be measured in a urine sample (obtained after digital rectal examination). A specific molecular test is used that compares levels of PCA3 against a 'background' level. More research is needed before the true value of this test is known.

It also seems likely that tests will soon be developed that predict the behaviour of individual prostate cancers more accurately, which will make it easier for patients, their families and their doctors to decide which is the best treatment option. One of these, the Prolaris test is based on the analysis of 41 cell cycle genes and has just become available in the UK.

## New treatments

### Drug treatments

As new anti-androgens are developed, it is likely that they will be used at earlier stages of the disease when the cancer cells are more sensitive to the blocking of the action of testosterone. Research is also being carried out into several drugs that block the pathways of the growth factors that are necessary for the development and progression of prostate cancer. In order for a cancer to grow, it requires a new blood supply. Drugs that block the growth of this blood supply have anti-cancer potential. These so-called angiogenesis inhibitors are currently being tested for activity against prostate cancer. Many of these new approaches offer the possibility of fewer side effects and greater effectiveness.

### Immunotherapy

Work on harnessing the immune system to counter prostate cancer may eventually make it possible to vaccinate men at high risk of the disease or induce an immune response against established disease.

An example is Provenge (common name sipuleucel-T). With this treatment, a patient's immune cells are mixed with a protein designed to produce an immune response in the body to prostate cancer. The immune cells are activated when mixed with this protein, and the activated cells are then infused into the patient's body. To date, this treatment is available only in the USA, and costs around $165,000 (£100,000) for a month-long course of treatment per patient. Other immunotherapies are in the pipeline, and hopefully will be considerably cheaper.

## Gene therapy

Spectacular advances in molecular biology have made the prospect of gene therapy an imminent reality. In the not too distant future, it may be possible to 'turn off' the oncogenes that induce cancer and 'turn on' the protective tumour suppressor genes. New therapies will also be developed that selectively destroy prostate cancer by activating the in-built cell suicide system known as 'apoptosis'.

# BPH: its symptoms, diagnosis and treatment

**Nearly half of men over the age of 65 suffer either urinary symptoms or a reduced urinary flow due to benign prostatic hyperplasia (BPH). BPH is characterized by the benign (non-cancerous) overgrowth of prostate cells, with the effect that the central portion of the prostate progressively enlarges. The result is that the part of the urethra that is surrounded by the prostate becomes constricted. This reduces the urinary flow and the man finds that his urine stream becomes weaker and it is more difficult to empty his bladder. These symptoms may significantly impair quality of life.**

In response to the increasing obstruction, the muscular bladder wall thickens and becomes stronger. Consequently, the pressure inside the bladder needed to produce urinary flow has to increase to overcome the effect of the obstruction; this high pressure causes pouches or 'diverticula' to form. Less commonly, the raised pressure results in what is known as 'back pressure' on the kidneys, causing kidney problems. If BPH is not treated, either chronic urinary retention (characterized by a massively over-distended bladder) or acute urinary retention (the sudden inability to pass any urine, with painful over-distension of the bladder) may develop. In either situation, hospital admission, catheterization and, often, prostate surgery are usually required.

## Why do some men suffer more than others?

Recent work has clarified the risk factors linked to an increased likelihood of developing complications of BPH. The larger the prostate (as assessed by digital rectal examination and ultrasound), the greater the risk. Similarly, the risk is increased among those men with a PSA level above 1.4 ng/mL. In fact the higher the PSA (provided prostate cancer is not present), the greater the risk of urinary retention. Also more likely to develop complications are men whose urine tends to flow slowly and those who have a relatively large amount of urine left in their bladder after attempting to urinate. Although not all men suffer progressive deterioration, in the majority of cases the symptoms gradually become worse over time and eventually complications develop.

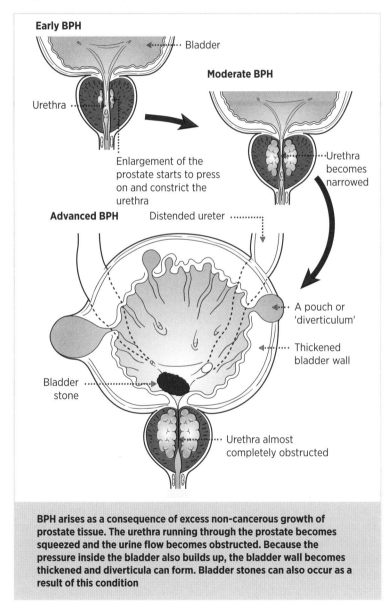

**BPH arises as a consequence of excess non-cancerous growth of prostate tissue. The urethra running through the prostate becomes squeezed and the urine flow becomes obstructed. Because the pressure inside the bladder also builds up, the bladder wall becomes thickened and diverticula can form. Bladder stones can also occur as a result of this condition**

# How is BPH diagnosed?

The spectrum of symptoms that are associated with BPH are known collectively as lower urinary tract symptoms (LUTS for short), and are outlined in the table overleaf.

The symptoms of BPH overlap with those of other conditions, so your initial examination should be thorough. Your doctor will question you about your general health and symptoms. In order to assess your symptoms systematically, you may be asked questions that relate to a scoring system (an example of this system is shown on page 24). Your doctor will also be concerned with how 'bothersome' you find your symptoms. Again, this can be approached in a systematic manner, and your responses can be scored. You will be asked about other conditions, such as diabetes and hypertension, and what medications you are taking.

## REFERRALS

**WHY YOU MAY BE REFERRED TO A SPECIALIST**

- Your symptoms appeared suddenly or are severe
- You have had repeated urinary infections
- You have passed blood in your urine
- Your PSA level is over 4 ng/mL (2.5 ng/mL in younger men) or has risen sharply
- Your GP thinks you may have a bladder stone
- The results from your blood tests suggest you might have kidney damage

## Physical examination and digital rectal examination

A digital rectal examination will be performed to give the doctor an idea of the size and consistency of your prostate (see page 22). He will also feel your abdomen to check whether your bladder is distended so that it can be felt (if it can, this is a sign that you may be retaining urine). Your doctor may also make an assessment of your nervous system, such as testing the muscle tone and sensation in the area around and between the scrotum and anus, as some disorders of the nervous system, such as Parkinson's disease or spinal cord problems, can give rise to urinary symptoms similar to those of BPH. Since high blood pressure (hypertension) is common, blood pressure may also be measured as part of a general health check.

## Urine test

As a urinary tract infection can cause symptoms such as an increased need to urinate, a urine sample will be checked for signs of bacterial infection or blood. The urine may also be tested routinely for the presence of sugar, which is a sign of diabetes. The urine may also be checked for malignant cells resulting from a bladder, ureteric or kidney cancer.

**LOWER URINARY TRACT SYMPTOMS ASSOCIATED WITH BPH**

- Hesitancy (when the urine flow stops and starts)
- A weak urine stream
- You need to strain to pass urine
- Urination takes a long time
- After urinating, you feel as though there is still some urine 'left behind' in the bladder
- When you get the urge to urinate, you feel you need to do so urgently
- Frequent trips to the toilet
- Getting up in the night to urinate
- When you get the urge to urinate, you leak a little urine
- A sudden or slowly building inability to urinate

## Blood tests

A very small proportion of men have kidney problems as a consequence of their BPH. By assessing the amount of a substance called creatinine in the blood, your doctor will be able to check whether your kidneys are affected by back pressure (see page 83).

Your blood sugar level may also be tested to check that you do not have diabetes, as this can be a cause of frequent urination.

The amount of PSA may also be measured. You might already have read about this in earlier chapters on prostate cancer. PSA is a marker that indicates damage to the prostate, often arising as a result of prostate cancer, but sometimes as a result of BPH. In fact, the larger your prostate, the higher your PSA tends to be. If your PSA level is raised, it may be recommended that you have a prostate biopsy so that prostate cancer can be excluded (see pages 27-28). As already mentioned, your PSA level also gives a rough indication of your prostate size, and this can influence the risk of you developing urinary retention and provide information about the likely success of various medical treatment options.

## Urine flow tests (or uroflowmetry)
By measuring the speed of your urine output over time, your urologist can get some useful information about your urine flow. For this test you will have to urinate (privately) into the bowl of a specialized piece of medical equipment known as a flow meter.

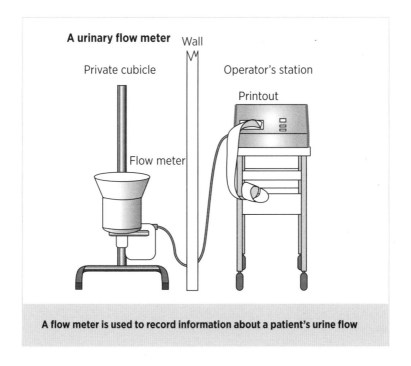

A flow meter is used to record information about a patient's urine flow

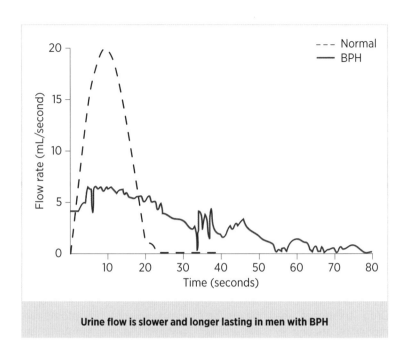

**Urine flow is slower and longer lasting in men with BPH**

## Ultrasound to measure urine left in the bladder

Ultrasound can give your doctor an idea of how severe the obstruction is and how well you might respond to certain types of treatment. The procedure is very similar to that used for pregnant women. High-frequency sound waves are emitted from a probe that is passed over your abdomen, and the echoes build up a picture that can be seen on a computer monitor.

## Less common tests

Depending on the results of the tests already described, your urologist may want to perform some further tests.

### Urodynamic measurements

Urodynamic measurements are made using a small catheter that is inserted up through the urethra, via your penis, into the bladder. By measuring the pressure within your bladder, your urologist can deduce whether your symptoms are due to obstruction from BPH or are the result of the bladder itself not working properly. This test is uncomfortable rather than painful, and takes around 20 minutes. You should drink extra fluids after the test to reduce the risk of subsequent urinary infection.

### Transrectal ultrasonography (TRUS)

Transrectal ultrasonography (TRUS) is used to visualize the prostate, measure its proportions and help guide a biopsy needle when there is a possibility of prostate cancer. The procedure is described fully on pages 26-28.

## Treatment

BPH is most commonly treated with either drugs or surgery. Some men with mild symptoms opt for active surveillance, which involves monitoring their condition so that any worsening can be quickly spotted and treated. There are also several 'minimally invasive' alternatives.

### Drug treatment

Drug treatment may be recommended if your symptoms are moderate, though it may also be beneficial if your symptoms are severe. Certain complications of BPH, such as kidney problems, urinary retention or bladder stones, make surgery a more appropriate option. There are two main classes of drug that are prescribed for BPH:

- alpha-blockers
- 5-alpha-reductase inhibitors.

### Alpha-blockers

Alpha-blockers such as tamsulosin work by helping to relax the muscles at the neck of the bladder and in the prostate. By reducing the pressure on the urethra, they help to overcome the obstruction and so increase the flow of urine. Results available from studies to date indicate that up to 60% of men find that their symptoms improve significantly within the first 2-3 weeks of treatment with an alpha-blocker.

This type of drug does not cure BPH, but simply helps to alleviate some of the symptoms. You may still develop complications at a later date and you may still need surgery eventually. The most commonly occurring drug side effects are tiredness, dizziness and headache, which affect around one in ten men.

### 5-alpha-reductase inhibitors

5-alpha-reductase inhibitors work by blocking the conversion of testosterone to another substance, DHT (dihydrotestosterone), that is known to have a key role in prostate growth. To date, most information is available on the 5-alpha-reductase inhibitor Proscar (finasteride); a newer agent, Avodart (dutasteride) is also now available. Unlike alpha-blockers, Proscar and Avodart do appear to be able to reverse the condition to some extent, particularly if the prostate is significantly enlarged, so their use may reduce the likelihood that you will develop acute urinary retention and eventually require surgery. These drugs also seem to work better in patients with larger prostate glands, but it can take 6 months or so for them to be effective. Importantly, they do reduce the PSA value by around 50% so this should be taken into account when monitoring for prostate cancer; one way to do this is to double the PSA value obtained when a patient is taking either Avodart or Proscar.

The main side effects of these agents are a reduced sex drive and difficulty in maintaining/achieving an erection; these appear to affect around 3–5 men in every 100. There is also a small chance of about 1% or less that you might experience tenderness and swelling around the nipples. These symptoms usually disappear if treatment is stopped. Be aware that crushed or broken Proscar or Avodart tablets should not be handled by a woman who is pregnant or who is planning a pregnancy, as there is a risk that they could cause problems to a developing baby.

### Combination therapy

Combination therapy with an alpha-blocker and a 5-alpha-reductase inhibitor has been shown to be more effective than either agent used alone in preventing the worsening of the symptoms of BPH or the development of complications, such as acute retention or the need for surgery. However, the increased cost and additional side effects have to be weighed against these benefits. Patients most likely to respond to combination therapy are those with both a large prostate and severe symptoms. A combination of dutasteride and tamsulosin in one pill, known as Combodart, has recently been released, and patients often find this to be more convenient. Side effects are a combination of those seen with the individual medications.

*Other medical strategies*

Other medical strategies for relief of urgency in BPH include anticholinergic agents like Detrusitol XL (tolterodine), Vesicare (solifenacin) and Toviaz (fesoterodine) to control urinary urgency and frequency. However, these agents carry a small risk of precipitating acute retention of urine in men with severe obstruction and may also result in a dry mouth.

Recently Cialis (tadalatil) at a dose of 5mg/day has been licensed for the treatment of BPH. It is especially useful in men who also complain of erectile dysfunction as it improves the rigidity of erections.

In patients who are particularly troubled by the need to pass urine during the night (nocturia), vasopressin analogues such as Desmospray or Desmotabs (desmopressin) last thing at night, used in addition to fluid restriction in the evenings, can be quite effective. These drugs work by reducing the amount of urine produced by the kidneys for 6–8 hours.

## Surgery

There are a number of surgical options for BPH:

- transurethral resection of the prostate (TURP)
- transurethral incision of the prostate (TUIP)
- open prostatectomy
- laser prostatectomy.

*TURP*

TURP is still the 'gold-standard' operation for men who have not responded to medical therapy or who have developed complications such as complete retention of urine, and is usually carried out under a general anaesthetic. It involves passing an instrument up through the penis, and then using it to cut the middle out of the enlarged prostate, piecemeal (see diagram below). A catheter is passed through the urethra into the bladder at the end of the operation to drain off the urine. This is left in place for a couple of days. A normal hospital stay following TURP is 3 or 4 days, but you should try to rest as much as possible for a few weeks afterwards to minimize the risk of secondary complications such as bleeding that may occur 10–12 days after the original operation.

After the operation, you may find that you experience an urgent need to urinate and/or a burning sensation when you pass urine. This should disappear within a few weeks. You may also notice some blood in your urine. This is normal, but if it is particularly heavy or persists for more than a few weeks, or if you notice some blood clots, drink extra fluids and contact your doctor.

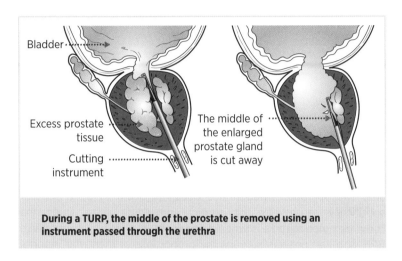

**During a TURP, the middle of the prostate is removed using an instrument passed through the urethra**

The most common side effect is a phenomenon known as retrograde ejaculation – where semen passes into the bladder during orgasm, rather than out through the penis (see diagram below). You then pass the semen mixed with urine the next time you urinate. This is not harmful and, providing that they know about this potential side effect before undergoing the surgery, most men do not find it bothersome. However, retrograde ejaculation will almost certainly reduce your fertility, though it does not make you reliably sterile.

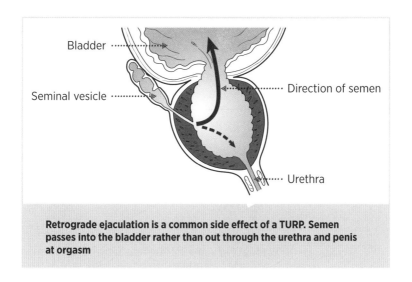

**Retrograde ejaculation is a common side effect of a TURP. Semen passes into the bladder rather than out through the urethra and penis at orgasm**

A few men complain of an inability to achieve or maintain an erection after the operation, though this does not seem to be a problem specifically caused by this surgical procedure. In a study that compared men with BPH who had undergone a TURP with men with BPH who had not had surgery, the proportions of men who reported erectile problems were similar. Some were even improved by surgery.

Some men notice some incontinence after a TURP – if you find that you are leaking urine slightly, talk to your doctor. This problem nearly always resolves completely with time, but if it persists further investigation and treatment may be warranted.

An operation under general anaesthetic always carries some small risks, as occasionally an individual reacts badly to anaesthesia. There is also a chance of significant blood loss and the subsequent need for a transfusion. In the post-operative period, there may be problems with catheter blockage or bleeding after the catheter has been removed. These problems are relatively unusual with a TURP, however, and the outcome is usually good. Narrowing of the urethra, a so-called 'urethral stricture', may also develop, but can usually be resolved with a further minor procedure.

When a TURP is performed the prostate tissue removed is sent to the pathology laboratory for analysis. In most cases the results come back confirming benign prostatic hyperplasia (BPH); however, in around one case in ten, a small amount of prostate cancer is identified. Small areas of prostate cancer may not require active treatment, but careful follow-up is indicated, with scanning and biopsy of the remaining prostate tissue, which may harbour some residual cancer tissue.

A new device has come on to the market that helps to reduce bleeding during TURP. Known as the 'button electrode', it vaporizes the prostate. However, this means that there is no tissue available to check for cancer.

### TUIP

TUIP (also known as 'bladder neck incision') is appropriate for the man who is experiencing obstruction problems but who has a relatively small prostate. It is quite quick to perform, taking only around 20 minutes, but you will still be given a general or spinal anaesthetic. As with a TURP, an instrument will be passed up through the penis, but rather than removing a portion of the prostate, one or two small cuts are made in the neck of the bladder and in the prostate. These have the effect of reducing the obstruction and allowing the bladder neck to spring apart. As with a TURP, you will be catheterized at the end of the operation to allow urine to drain away freely. The catheter will be removed after around 24–48 hours, and you will be able to leave hospital after a couple of days. For the next week or so, you should take things easy.

The chance that you will experience a side effect following a TUIP is lower than following a TURP. Retrograde ejaculation (see page 93), for example, affects a much lower proportion of men after the operation (1 in 10 compared with 8 in 10). There is a risk that symptoms will return after the operation (see table on page 97); if this happens, then it is likely that you will need a TURP.

Again, as the operation is performed using a general anaesthetic, there is a small risk of anaesthetic-related complications as well as surgical problems such as post-operative bleeding.

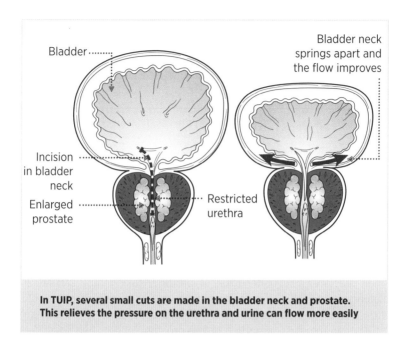

**In TUIP, several small cuts are made in the bladder neck and prostate. This relieves the pressure on the urethra and urine can flow more easily**

### Open prostatectomy

Open prostatectomy is only really appropriate for the man whose prostate is very large (more than 100 grams) or who has large bladder stones. It is a more complex procedure than a TURP, and complications afterwards are somewhat more likely.

The surgeon gains access to the prostate through a horizontal incision made in the lower abdomen. Through a cut made either in the prostate or bladder, the surgeon is then able to remove the central part of the prostate. A catheter is inserted to drain the bladder, and this is left in place for 3 or 4 days.

Because this is relatively major surgery, you will usually need to stay in hospital for 5-7 days. Even when you go home, you are advised to rest for up to 6 weeks, and you should avoid lifting anything heavy for several months. The operation will leave a scar.

An open prostatectomy can also result in retrograde ejaculation (see pages 93), with about seven in ten men being affected; some men also find it difficult to achieve/maintain an erection (around two men in ten). Other problems associated with surgery of this type are similar to those described in the section on TURP.

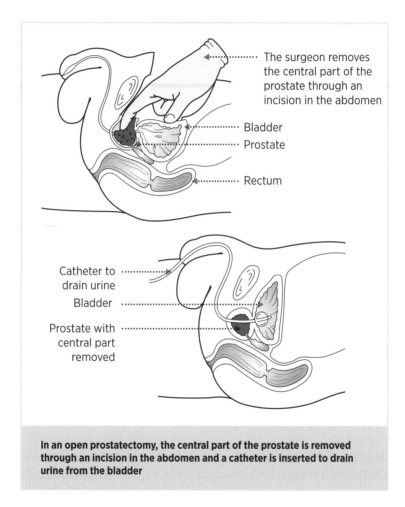

In an open prostatectomy, the central part of the prostate is removed through an incision in the abdomen and a catheter is inserted to drain urine from the bladder

Useful information comparing the outcomes following each surgical procedure is presented in the table below.

## OUTCOME AFTER THE THREE MAIN SURGICAL OPTIONS FOR TREATMENT OF BPH

|  | TURP | TUIP | Open prostatectomy |
|---|---|---|---|
| Likelihood that symptoms will improve | 90% | 80% | 98% |
| Usual reduction in symptom score (see page 24) | 85% | 73% | 79% |
| Likelihood that you will need further surgery within 8 years | 16-20% | Over 20% | 10% |

### Laser therapy

As has already been said, minimally invasive treatments are relatively new. While greeted with enthusiasm by some urologists – and many patients – it has to be said that, currently, some minimally invasive techniques do not always work as well (or as definitively) as the more traditional treatments. One that particularly deserves a mention is laser therapy.

Laser therapy is carried out under general anaesthetic. A laser probe is inserted up through the penis, and the laser energy it generates destroys some of the prostate tissue. The 'GreenLight' laser vaporizes the prostate so that no tissue is available for examination under the microscope; higher power 180 watt GreenLight lasers have recently been introduced and are becoming popular.

Another laser technique – holmium laser enucleation of the prostate or HoLEP – allows the prostate to be cut away, and the pieces are then broken down into a paste and removed from the bladder by suction.

This may then be analyzed under the microscope. Bleeding is minimal after laser therapy, but catheterization is usually necessary for a short time afterwards. For a while, a burning sensation may be experienced on passing urine, which may be prolonged and quite troublesome. If this persists, a urine specimen should be sent to the laboratory for culture to rule out a urinary tract infection. The laser techniques are especially suitable for patients taking anticoagulants because they generally cause virtually no bleeding; they are becoming more popular as more and more specialized laser machines become available.

As with any type of surgery, experience with the technique improves the results, so you should enquire how many operations the team has performed.

## CASE STUDY

**EDWARD,** a 58-year-old designer, noticed a gradual diminution of his urinary stream and a feeling of incomplete bladder emptying. He was also getting up two or three times per night to pass urine. He consulted his GP who examined him and found a smoothly enlarged prostate, and checked his PSA and found it normal at 2.6 ng/mL. He was started on the alpha-blocker Flowmaxtra (tamsulosin), 0.4 mg/day, and the 5-alpha-reductase inhibitor Avodart (dutasteride), 0.5 mg/day, as combination therapy, and some improvement was noted.

After 8 months, however, he returned to his doctor and requested a referral to a urologist as he was still getting up at night and consequently feeling tired during the day. He was seen promptly, and a flow test and bladder ultrasound confirmed that he was still obstructed with a slow flow and incomplete bladder empting. A TURP was advised and this was accomplished uneventfully. Pathological examination of the prostate tissue removed confirmed BPH. Subsequently Edward passed urine less frequently with a much improved flow. Although he did develop retrograde ejaculation, this has not proved bothersome to him and he is happy with the outcome of treatment.

### Active surveillance

Active surveillance (no immediate treatment but regular monitoring) may be recommended if your symptoms are mild or if you are not too troubled by them. Your doctor will advise you about small changes that you can make to your lifestyle that might help: for example, try not to drink large volumes of fluid in the evenings. Tea, coffee and alcohol can worsen symptoms. Be sure to inform your doctor if symptoms worsen.

### Plant extracts (phytotherapy)

There is an increasing range of plant extracts on the market that supposedly alleviate BPH, and many claims have been made as to their effectiveness. However, scientific data from properly conducted long-term studies to support their safety and usefulness are limited.

A recent report on saw palmetto suggested that it was no more effective than inactive placebo; however, some patients swear by it. Nevertheless, phytotherapy almost certainly does no harm, it is relatively cheap and most urologists do not actively discourage its use.

## Prevention

Although we have still to identify the fundamental steps in the development of BPH, we know that testosterone is certainly involved in some way and it is likely that the female hormone oestrogen also has a role. Epidemiological data suggest that men in the Far East are protected, to some extent, against the risks of BPH by minute amounts of oestrogen-like substances in the food that they eat (for example, soya contains the phytoestrogen genistein). This raises the question as to whether dietary supplements taken regularly by men in Europe, the USA and elsewhere could protect against the risk of this disease. Long-term studies involving many men are needed to confirm or refute this.

# Prostatitis: The painful prostate

Prostatitis literally means 'inflammation of the prostate'. In fact, by no means every patient suffering from prostatitis actually has an inflamed prostate, so the name is rather misleading. Surprisingly, in the UK, the condition accounts for almost one-quarter of all consultations with a urologist.

Patients with prostatitis often suffer pain and discomfort in the area around and between the anus and scrotum, and just above the pubic bone. Men with the condition may have to urinate frequently and this can be very inconvenient. There may also be a burning sensation at the time of urination and/or a degree of discomfort during or after ejaculation (see below for a more complete list of symptoms).

## SYMPTOMS

### SYMPTOMS OF PROSTATITIS

- Chills and fever
- Pain in:
    - lower back (may be particularly painful after sex)
    - between the scrotum and rectum
    - penis
    - prostate (felt as lower abdominal pain and pain in the area between the scrotum and anus)
    - testicles
    - rectum
    - inner thighs
- Pain/difficulty in passing urine
- A need for frequent urination

Although prostatitis is often considered to be the result of a bacterial infection in the prostate, inflammation, when present, more commonly occurs spontaneously. Some studies suggest that in the absence of infection, inflammation may result from urine being forced backwards up the prostatic ducts at the time of urination. Recently, the question

**CLASSIFICATION OF PROSTATITIS (NATIONAL INSTITUTES OF HEALTH, USA)**

**Category I - acute bacterial prostatitis**
Acute bacterial prostatitis is usually caused by a bacterial urinary tract infection and is the least common, but most severe, form of prostatitis. Sufferers can feel generally unwell.

**Category II – chronic bacterial prostatitis**
Chronic bacterial prostatitis is a chronic or recurrent infection of the prostate that may be present for several years before any symptoms develop. The symptoms may be less aggressive, but are recurring.

**Category III – chronic (abacterial) prostatitis**
- IIIA: inflammatory (chronic pelvic pain syndrome)
- IIIB: non-inflammatory (prostatodynia)

Chronic (abacterial) prostatitis is the most common form of prostatitis and is a recurring (relapsing) condition. It is difficult to pinpoint a specific cause for this condition, but it is thought that an abnormal immune system reaction or a chemical reaction to urine flowing backwards into the prostatic ducts could play a major role. Earlier suggestions that some sexually transmitted infections, such as chlamydia or mycoplasma, might be responsible have been disproved.

**Category IV – asymptomatic inflammatory prostatitis**
Some men feel discomfort that appears to come from their prostate or the surrounding area without any infection being present. Although there are many theories as to its cause, more research is required, as the inflammation could possibly lead to more serious problems, such as prostate cancer.

has been raised as to whether, in the long run, inflammation can lead on eventually to prostate cancer. This is a possibility because the inflammatory cells that infiltrate the prostate release chemicals that cause oxidative stress and thereby damage DNA in prostate cells.

Even when infection is the source of inflammation, it may be difficult to eradicate because the bacteria responsible tend to be inaccessible to antibiotics. This is because they usually lurk deep inside the prostate (for example, the bacteria may be inside the tiny stones that form in the prostatic ducts).

## Risk factors

Men who have an increased risk of bacterial prostatitis include those who have a previous history of the problem or long-term catheterization and those who have urinary tract infections that remain untreated. Unprotected anal intercourse may also spark off a bout of prostatic infection.

Prostatitis most commonly affects men in the age range 30–50 years, but a man of any age can be affected. In fact, most men afflicted with non-bacterial prostatitis have no identifiable risk factors.

## Tests

Because prostatitis is often the result of a bacterial infection, your doctor will usually want to check a sample of your urine and prostatic secretions for bacteria (the sample will be sent to the laboratory for analysis, so you will not get the results straight away).

### Obtaining a sample of prostatic secretions

You will be asked to pass urine and provide a sample. Your prostate will then be massaged so that secretions are released, which will be collected from the urethra into a sterile pot. Although this process is unquestionably a little uncomfortable, it is not actually painful (see the diagram on page 105 for an idea of what's involved). Finally, a second urine sample will be collected. If there is a bacterial infection, bacteria can be grown on an agar plate in the laboratory from cultures of the prostatic secretions and the second urine sample. This method also allows the specific type of bacteria responsible to be identified, and an appropriate antibiotic to be prescribed.

### Other tests

Depending on your symptoms, your doctor may also check that you do not have BPH or prostate cancer – the tests that may be performed are discussed on pages 22 and 23. Remember that prostatitis, particularly when the inflammation is severe, may sometimes cause a temporary increase in blood PSA level (see pages 15 – 16). Prostatitis can also cause blood flow in the prostate to become increased, and this can show up when a transrectal ultrasound study of the prostate is performed using what is known as a colour Doppler probe. If a prostatic abscess is suspected, a CT or MRI scan may be arranged to confirm or exclude a collection of pus in the prostate.

## CASE STUDY

**MATTHEW,** a banker aged 34, presented to his urologist with a long history of intermittent pelvic and perineal pain, together with bouts of frequency of urination. Several courses of antibiotics including ampicillin and oxytetracycline had resulted in only temporary improvement. He was eventually referred to a urologist who found a tender prostate on examination and sent some prostatic secretions off for culture. These came back showing inflammatory cells, but no bacterial growth. A 6-week course of Ciproxin (ciprofloxacin) and Brufen (ibuprofen) was prescribed and lifestyle modification advised. The medications, in addition to the running, swimming and improved diet, resulted in a marked improvement. Matthew has been warned, however, that relapse is possible and is therefore resolved to maintain the positive changes in his lifestyle.

## Treatment

If a bacterial infection is the cause of your symptoms, you will be prescribed a course of appropriate antibiotics. You may need to take these for a relatively long period, often 4–6 weeks, and it is very important that you complete the course (the diagram on page 106 explains why this is so). You may also be prescribed an anti-inflammatory drug, such as Voltarol (diclofenac), to reduce the inflammation in the prostate. Be aware that these drugs can cause indigestion and bleeding from the stomach, so you should always take them with a meal.

If bacteria are not demonstrably present, you may be given an anti-inflammatory drug in isolation. However, antibiotics can still also be helpful in these circumstances, perhaps because the cultures do not tell the whole story, as the bacteria may be lurking within prostate stones or elsewhere deep within the gland.

Some clinicians believe that frequent vigorous prostate massage by your urologist can also be beneficial, though this remains unproven.

Prostatitis, though troublesome, is not a life-threatening condition and is not proven to be a precursor to either BPH or prostate cancer. You may find that, over the years, the prostatitis returns from time to time (particularly if the condition has a non-bacterial cause), but your doctor should be able to help alleviate the symptoms quite effectively,

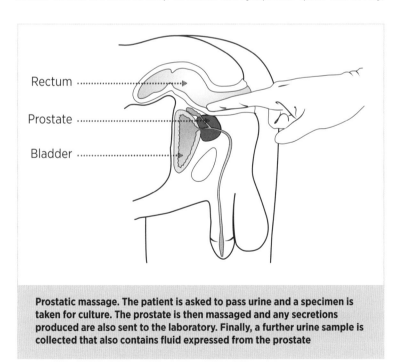

Prostatic massage. The patient is asked to pass urine and a specimen is taken for culture. The prostate is then massaged and any secretions produced are also sent to the laboratory. Finally, a further urine sample is collected that also contains fluid expressed from the prostate

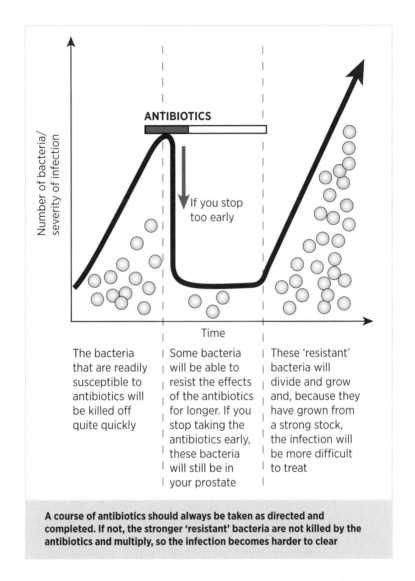

**A course of antibiotics should always be taken as directed and completed. If not, the stronger 'resistant' bacteria are not killed by the antibiotics and multiply, so the infection becomes harder to clear**

so do not suffer in silence. The condition can be frustrating for patient and doctor alike; however, there is some evidence that lifestyle improvements, particularly stress reduction, dietary modification and fitness enhancement, can increase natural immunity and reduce the risk of relapse.

## Prostatic abscess

Occasionally prostatitis with a bacterial cause can lead to the formation of an abscess within the prostate itself. If this is the case, your doctor may need to take a sample of fluid from the abscess and will do so using ultrasound for guidance, as in prostate biopsy (see pages 26–27). The sample can then be checked to see what type of bacteria has caused the infection, so that appropriate antibiotics can be given.

Abscesses sometimes have to be drained, which involves passing an instrument up through the penis under anaesthetic, making a small cut through to the abscess and then 'nicking' the abscess to allow pus to drain out. A course of antibiotics and a period of catheterization are also often necessary, as passing urine can be difficult or impossible because of the associated swelling.

## Preventing prostatitis

Prostatitis is the affliction of the prostate about which we know least. So, not surprisingly, we currently have little idea how to prevent the problem. The best advice at present is to avoid the risk factors for prostatitis where possible (for example, if you have symptoms of a urinary tract infection, such as a burning sensation when urinating, or cloudy smelly urine, visit your doctor and complete any prescribed courses of antibiotics). Also, maintain a healthy lifestyle, eat a diet low in saturated fats, take plenty of exercise and always be vigilant about avoiding a risk of infection, especially if you practise anal intercourse.

## Pain but no inflammation: prostatodynia

Some men feel pain that appears to come from their prostate or the surrounding area, but investigation does not reveal any inflammation or infection. What causes this condition, which is referred to as prostatodynia or pelvic pain syndrome, is not known, though it may result from spasm of the pelvic muscles brought on by stress and anxiety. Depending on your symptoms, you may be given alpha-blockers (see page 89), the PD5 inhibitor tadalafil, or the muscle relaxant diazepam. Much more research is needed into this disorder to identify the underlying cause of the problem, to improve the cure rate and enhance quality of life. Reducing stress levels and maintaining a healthy life-style may be helpful.

# Medications commonly used to treat prostate disorders

## Alpha-blockers

Alpha-blockers are used in BPH to help relax the muscles in the bladder and the prostate, which in turn helps to reduce the pressure on the urethra. Possible side effects include headaches, dizziness and nasal stuffiness.

| Alpha-blocker | Brand name* | Recommended dose** |
|---|---|---|
| Indoramin | Doralese | 20 mg twice daily |
| Doxazosin | Cardura XL | 4 mg once daily |
| Alfuzosin | Xatral XL | 10 mg once daily |
| Terazosin | Hytrin | 1 mg once daily |
| Tamsulosin | Flomaxtra XL | 0.4 mg once daily |

*Not all products are listed.
**Usual starting dose given in most instances; your doctor may adjust your dose over time.

## 5-alpha-reductase inhibitors

5-alpha-reductase inhibitors are used in BPH to block the conversion of testosterone to another substance, DHT, which appears to stimulate overgrowth of prostate tissue. Possible side effects include loss of libido, reduced erections and rarely minor breast enlargement.

| 5-alpha-reductase inhibitor | Brand name* | Recommended dose** |
|---|---|---|
| Finasteride | Proscar | 5 mg once daily |
| Dutasteride | Avodart | 0.5 mg once daily |

*Not all products are listed.
**Usual starting dose given in most instances; your doctor may adjust your dose over time.

## LHRH analogues

LHRH analogues are used in prostate cancer to 'switch off' testosterone production. They can also result in loss of sex drive and in hot flushes.

| LHRH analogue | Brand name* | Recommended dose** |
|---|---|---|
| Goserelin | Zoladex | 3.6 mg/month or 10.8 mg every 3 months |
| Leuprorelin | Prostap | 3.75 mg/month or 11.25 mg every 3 months |
| Triptorelin | Decapeptyl SR | 3 mg/month or 11.25 mg every 3 months |

*Not all products are listed.
**Usual starting dose given in most instances; your doctor may adjust your dose over time.

## LHRH antagonist

An LHRH antagonist blocks the process that leads to testosterone production. The side effects are similar to those experienced with LHRH analogues, but the flare reaction does not occur with an LHRH antagonist (see page 66). Allergic reactions around the injection site are common.

| LHRH antagonist | Brand name | Recommended dose |
|---|---|---|
| Degarelix | Firmagon | 240 mg (given as two injections of 120 mg), then 80 mg every 28 days |

## Anti-androgens

Anti-androgens are used in prostate cancer to block the action of testosterone. Side effects can include breast tenderness and enlargement.

| Anti-androgen | Brand name* | Recommended dose** |
|---|---|---|
| Cyproterone acetate | Cyprostat | 200 mg in two to three divided doses daily*** |
| Flutamide | None | 250 mg three times daily |
| Bicalutamide | Casodex | 50–150 mg once daily |

*Not all products are listed.
**Usual starting dose given in most instances; your doctor may adjust your dose over time.
***Lower doses may be used to help control hot flushes.

## Anticholinergics and anti-spasmodics

Anticholinergics and anti-spasmodics are used in BPH to treat irritative symptoms including urinary frequency and urgency. They may cause a dry mouth and blurred vision.

| Anticholinergic/anti-spasmodic | Brand name | Recommended dose |
|---|---|---|
| Flavoxate | Uripass | 200 mg three times daily |
| Oxybutynin | Cystrin | 5 mg two to three times daily |
| | Lyrinel XL | 5 mg once daily |
| Propiverine | Detrunorm | 15 mg one to three times daily |
| Tolterodine | Detrusitol XL | 4 mg once daily |
| Trospium | Regurin | 20 mg twice daily |
| Solifenacin | Vesicare | 5–10 mg once daily |
| Fesoterodine | Toviaz | 4–8 mg once daily |

## Vasopressin analogues

Vasopressin analogues are used in BPH to reduce the amount of urine produced at night, thus reducing the need to get out of bed to pass urine. They should not be taken with large amounts of fluid and are not advised for men older than 65 years of age.

| Vasopressin analogue | Brand name | Recommended dose |
|---|---|---|
| Desmopressin | DDVAP nasal | 10–20 µg at bedtime |
| | Desmospray | 10–20 µg at bedtime |
| | Desmotabs | 200 µg at bedtime |

## Other drugs

Anti-inflammatory agents may be given to control pain and inflammation; for example, diclofenac (brand name Voltarol SR).

A wide range of antibiotics may be given to treat infection, such as ciprofloxacin (brand name Ciproxin) and co-amoxiclav (brand name Augmentin). Diazepam (Valium) is sometimes used as a muscle relaxant and a means of relieving anxiety. Cialis (tadatalil) 5 mg/day is used to treat men with lower urinary tract symptoms due to benign prostate hyperplasia (BPH) and associated erectile dysfunction. It may also be useful in men with prostatodynia and chronic abacterial prostatitis, but at the moment this is experimental.

# Further information and support

When seeking further information and support, it is important to choose your source carefully. The internet, in particular, is a popular source of material, but sites are unregulated and much of the information is unvalidated and sometimes frankly promotional. The sources listed below are just some of the myriad books, websites and charities out there, but they should provide further sources of help and information about all aspects of prostate disorders, their treatments and their effects. Although the sources have been put into different groups (and some are listed more than once), there is considerable overlap and the headings are just suggestions as to the best place to look for information first.

## General health and lifestyle
- The NHS provides guidance on portion sizes for fruit and vegetables at www.5aday.nhs.uk, or telephone the NHS Response line on 08701 555455 and ask for a leaflet with details of typical portion sizes.
- Men's Health Forum (www.menshealthforum.org.uk) adopts a number of strategies to improve the health of men and men's health services. One of these is a website called MALEHEALTH (www.malehealth.co.uk), which provides essential, accurate and easy-to-use information about the key health problems that affect men. It also includes an online health check.
- Health of Men (www.healthofmen.com) is a 5-year Big Lottery Fund initiative that provides "quick clear health information for boys and men". This includes advice about all aspects of lifestyle, such as eating, exercise and dealing with stress, as well as other men's health issues.
- NHS Direct gives a wide range of information about health, conditions, treatments and local services. It can be accessed either by telephone (helpline 0845 4647), via the internet (www.nhsdirect.nhs.uk) or, if you have digital satellite television, via the NHS Direct Interactive service.

- www.embarrassingproblems.com is a website that covers health problems that are difficult to discuss with anyone.
- SAGA Health (www.saga.co.uk/health_news) provides broadly based health information for the over-50s. It also includes sections on medicines and supplements, as well as complementary medicine.
- The website for the journal Trends in Urology and Men's Health (www.trendsinurology.com) has a 'patient resources' section from which many helpful leaflets and information sheets can be downloaded.

## Prostate disorders

- Prostate Cancer UK (www.prostatecanceruk.org) provides clear concise and up-to-date information. Phone 0800 074 8383 to speak to their specialist nurses.
- UK Prostate Link (www.prostate-link.org.uk) provides a searchable database of quality-assessed links to prostate cancer information on the internet. It also has links to a number of sites featuring personal experiences.
- Macmillan Cancer Support (www.macmillan.org.uk) provides comprehensive information about all aspects of prostate cancer from diagnosis to the latest clinical trials. Helpline (staffed by specialist cancer nurses): 0808 808 0000.
- The British Prostatitis Support Association (www.bps-assoc.org.uk) is a web-based organization that offers information and support to sufferers of prostatitis, male chronic pelvic pain syndrome and interstitial cystitis.
- Health Press publishes a number of books by Professor Roger Kirby and others covering prostate disorders. Patient Pictures: Prostatic Diseases and their Treatments provides a simple guide to the most common procedures used to treat prostate problems. Fast Facts: Prostate Cancer and Fast Facts: Benign Prostatic Hyperplasia, although written for doctors, are also read by patients wanting more detailed information. Order from www.fastfacts.com or telephone 01752 202 301.
- National Institute for Health and Clinical Excellence (NICE) has an excellent website with good information on prostate cancer and BPH.

## Treatment
- Besttreatments (www.besttreatments.co.uk) is run by the British Medical Journal with the aim of helping you make better health decisions. It looks at all the best research evidence and decides how well treatments work.
- The National Institute for Health and Clinical Excellence (NICE) produces evidence-based guidance for the NHS. Enter 'prostate' in the search box on their website (www.nice.org.uk) to see the latest guidance.
- The American Cancer Society (www.cancer.org) provides an online decision tool to help you understand the treatment options for prostate cancer and the possible side effects.
- Bandolier (www.medicine.ox.ac.uk/bandolier) is an independent journal about evidence-based healthcare, written by Oxford scientists. They find information about evidence of effectiveness (or lack of it), and put the results forward as simple bullet points of those things that worked and those that did not.
- The National Cancer Institute in the USA (www.cancer.gov/prostate), as you might expect, provides comprehensive information about prostate cancer, but also includes information and current news about clinical trials and trial-related data.

## Continence
- The Bladder and Bowel Foundation (www.bladderandbowelfoundation.org) offers information, advice and expertise about bladder and bowel problems, no matter how small. Their helpline is staffed by specialist nurses who will be able to give you the information and advice you need, and also tell you where to find your local NHS specialist continence service: 0845 345 0165.

## Sexuality
- The Sexual Dysfunction Association (www.sda.uk.net) has a wealth of advice on male and female sexual problems. Helpline: 020 7486 7262.
- Sorted in 10 (www.sortedin10.co.uk) offers information about erectile dysfunction and gives advice on approaching your doctor and treatments available, as well as advice for partners on dealing with impotence.

## Practical support
- Macmillan Cancer Support (www.macmillan.org.uk) provides information on practical issues via its website.
- Marie Curie Cancer Care (www.mariecurie.org.uk) also provides information and support. You can contact them by phone on 0800 716 146 or by email at supporter.services@mariecurie.org.uk.
- Your local library may also be able to provide you with details of services available from your local council.

## Support groups
- Websites such as Macmillan Cancer Support (www.macmillan.org.uk) have online communities where you can share experiences with others.
- Local support groups will enable you to meet others in a similar situation. Your GP surgery may be able to put you in touch with relevant groups.
- The American site PSA Rising (www.psa-rising.com) provides information and support for prostate cancer survivors. The site offers online forums and the latest news about prostate cancer and its treatment.

# Some medical terms explained

**Active surveillance:** careful follow-up of low-risk cancer with PSA testing, MRI and repeat biopsy, which allows treatment to be avoided or postponed.

**Adjuvant therapy:** a treatment that enhances the effectiveness of another therapy.

**Advanced:** cancer is described as advanced when it has spread beyond the site where it started. Prostate cancer is described as being locally advanced when it has invaded parts of the body around the prostate. When the cancer has begun to spread to more distant sites, such as the bones, it is no longer localized and so is referred to as advanced.

**Alpha-blockers:** one of the two types of drug usually prescribed for BPH. They work by helping to relax muscles in the bladder and prostate, which in turn helps to reduce the obstruction to the urinary tract caused by an enlarged prostate.

**5-alpha-reductase inhibitors:** one of the two types of drug usually prescribed for BPH. They work by blocking the conversion of testosterone to another substance, DHT (see page 109), which appears to stimulate overgrowth of prostate tissue.

**Angiogenesis:** the development of a blood supply. As a tumour grows, the formation of a blood supply is vital to cancer cells so that they can survive and divide. Researchers are currently developing drugs that could hinder angiogenesis, and so stop cancers growing.

**Anti-androgens:** drugs that may be prescribed to combat prostate cancer. They work by blocking the action of testosterone, which appears to stimulate the growth of prostate cancer.

**Anticholinergic agents:** drugs sometimes used to control urinary urgency and frequency associated with BPH.

**Benign:** non-cancerous; an area of unregulated tissue growth that does not have the capacity to invade surrounding healthy tissue or metastasize.

**Biopsy:** a sample of tissue taken from the body. Biopsies of prostate tissue are checked under a microscope for signs of cancer.

**Bone scan:** a means of seeing whether the cancer has spread to the bones. It involves injecting the patient with a radioactive material that then spreads around the body and then capturing the image a few hours later. The final pattern of distribution will highlight any areas where cancer may be developing.

**BPH (benign prostatic hyperplasia):** a non-cancerous condition that causes the prostate to become enlarged, which may lead to difficulty with urination.

**Brachytherapy:** a type of radiotherapy for prostate cancer in which radioactive pellets are implanted into the prostate.

**Cancerous:** refers to unregulated tissue growth that has developed the ability to invade surrounding healthy tissue.

**Catheter:** a narrow tube inserted into the penis and up into the bladder to drain urine away. A catheter may be inserted during an operation so that the bladder does not fill with urine while the surgeon is working on it. It may be left in place for some time afterwards so that the patient can pass urine while his urethra and bladder heal. Sometimes catheters are also inserted via the penis so that fluid can be passed into the bladder; for example, see 'Urodynamics'.

**Cavernous nerves:** the nerves involved in sexual arousal and erection that lie close to the prostate. They may be disturbed during radical prostatectomy.

**Cells:** tiny, specialized units from which the body is built. Healthy cells grow and divide as part of their normal lifecycle; in cancer, these processes get out of control because the usual mechanisms that keep them in check have broken down.

**Chemotherapy:** the use of drugs to destroy cancer cells.

**Conformal radiotherapy:** a type of radiotherapy that conforms to the shape of the prostate and thereby reduces the radiation dose to nearby tissues.

**Continence:** the ability to maintain control over bladder and bowel emptying.

**CT scanning:** a method of using sequential X-rays to build up a three-dimensional picture of the body. CT stands for 'computed tomography'.

**CyberKnife:** a robotically controlled form of giving radiation treatment to the prostate.

**Cystoscopy:** the use of a telescope, inserted through the penis, to examine the inside of the bladder.

**Da Vinci:** the brand name of the robot used by surgeons to facilitate robotic radical prostatectomy.

**DHT:** the male hormone testosterone can be converted in the body to DHT, which is thought to stimulate the growth of prostate tissue. DHT stands for 'dihydrotestosterone'.

**Differentiated (as in 'well, moderately well or poorly differentiated'):** a term used to describe healthy organized tissue. As cancer invades, the tissue structure becomes disorganized or de-differentiated, and looks less and less like normal tissue.

**Digital rectal examination:** a procedure that allows the doctor to assess the size and texture of the patient's prostate gland. It involves placing a finger into the patient's back passage (rectum) and feeling (palpating) the gland. It is sometimes referred to as a DRE.

**First-degree relative:** a close relative (parent, sibling or child). Prostate cancer may run in families, and more than 70 genes have now been identified that confer prostate cancer susceptibility.

**Gland:** a group of cells with the specialized function of making a particular fluid or secretion. The fluid made in the prostate gland mixes with the jelly-like sperm to make semen, which can then be ejaculated.

**Gleason score:** a number from 2 to 10 that is used as an indicator of how aggressive the patient's cancer is. The score is derived from an assessment (Gleason grade) of two areas of a sample (biopsy) of prostate tissue (e.g. 3 + 4 = 7).

**Grade:** how the prostate tissue appears under a microscope. The more aggressive the cancer, the less it looks like normal prostate tissue. The Gleason grading system uses a scale of 1–5, with 5 indicating the most aggressive-looking cancer.

**GreenLight laser prostatectomy:** a technique that vaporizes the obstructing prostate tissue to improve urinary flow with minimal risk of bleeding.

**HIFU (high-intensity focused ultrasound):** a new technique that focuses ultrasound waves on prostate cancer cells. More research is needed before it can be regarded as a mainstream treatment; it is currently available only as part of a clinical trial.

**Holmium laser enucleation of the prostate (HoLEP):** a technique that cuts away the prostate bloodlessly. The tissue is then cut into tiny pieces and removed from the bladder by suction.

**Hormones:** usually described as 'chemical messengers', these substances can influence processes at different sites in the body. Testosterone is a well-known hormone that influences many aspects of 'maleness'.

**Hormone-relapsed prostate cancer:** prostate cancer that has responded initially to hormone therapy but is beginning to grow again with a consequent rise in PSA values.

**Hormone therapy:** the use of drugs to block the stimulatory effects of testosterone on the growth of prostate tissue. Technically, orchidectomy can also be described as hormone therapy, as the testicles are removed so that testosterone is no longer produced.

**Immunotherapy:** Provenge (sipuleucel-T) therapy involves taking the patient's white blood cells and treating them to make them attack prostate cancer cells. It is not available in the UK at the time of writing.

**Impotence (or erectile dysfunction):** a state in which a rigid erection cannot be achieved and/or maintained.

**Intensity modulated radiotherapy (IMRT):** a type of radiotherapy that can target high doses of radiation to a very specific area of the body, thereby reducing the dose of radiation to normal tissues nearby.

**Laser prostatectomy:** a technique to treat obstruction due to BPH that involves almost no blood loss.

**LHRH analogues:** drugs used in hormone therapy for prostate cancer. They work by switching off testosterone production. LHRH stands for 'luteinizing hormone releasing hormone'.

**LHRH antagonist:** another type of drug used in hormone therapy. An LHRH antagonist blocks the production of testosterone without producing the temporary surge in male hormone production at the beginning of treatment that can occur with LHRH analogues.

**LUTS:** lower urinary tract symptoms. The term used to describe the range of symptoms associated with BPH.

**Lymph nodes:** these occur at intervals throughout the lymphatic system and act as filters, so cells such as cancer cells tend to accumulate at these points. A well known example of the lymph nodes (or 'glands') lie in the neck just below the jaw; these tend to become swollen during flu-type illnesses.

**Lymphatic system:** a network of vessels that drain fluid (lymph) from the body's organs so that it can be filtered and returned to the blood. It also works as part of the immune system.

**Malignant:** see Cancerous.

**Maximal androgen blockade:** the use of LHRH analogues and long-term anti-androgens to help slow the progression of prostate cancer.

**Metastases:** deposits of cancer distant from the site of the original cancer. A cancer has the ability to metastasize when cells can break off from the primary tumour and establish secondary tumours at other sites.

**MRI:** a means of building up a three-dimensional picture of the body using magnetic fields. MRI stands for 'magnetic resonance imaging'.

**Oncologist:** a doctor who specializes in the medical treatment of cancer.

**Open prostatectomy:** an operation for BPH that involves removing the central part of the prostate. Access is gained via an incision through the abdominal wall.

**Orchidectomy:** an operation for prostate cancer in which both testicles are removed from the scrotum so that testosterone production ceases.

**Palliative care:** this becomes important in the later stages of cancer where the aim of the medical team is to make the patient pain-free and as comfortable as possible.

**Pathologist:** a doctor who examines tissue samples microscopically to obtain information to help with diagnosis and treatment.

**PCA3:** prostate cancer antigen 3. A genetic marker for prostate cancer found in the urine after vigorous massage of the prostate gland.

**Pelvic pain syndrome:** see prostatodynia.

**Perineum:** the area around and between the scrotum and anus.

**Peripheral zone:** the part of the prostate gland in which prostate cancer usually starts to develop. It is also the part that usually becomes inflamed in prostatitis.

**Phytotherapy:** the use of plant extract to combat illness, such as BPH.

**PIN (prostatic intraepithelial neoplasia):** the earliest stage in uncontrolled cell growth. It is not cancer, but can be a forerunner to it.

**Plasma button prostatectomy:** a new form of prostate operation for BPH. The tissue is vaporized rather than cut away (resected), which reduces bleeding.

**Prostatodynia:** a state in which pain apparently comes from the prostate or surrounding area but there does not appear to be any inflammation or infection.

**PSA (prostate-specific antigen):** a substance made in the prostate gland that helps to liquefy the jelly-like sperm. If the prostate tissue becomes damaged or disrupted, as is particularly the case with prostate cancer, PSA leaks out into the bloodstream. As a consequence, blood levels of PSA tend to be higher among men with prostate cancer. A normal PSA value is usually taken as being below 4 ng/mL ('nanograms per millilitre'), but cancer can be present when values are lower than this. Conversely, a raised PSA is not always indicative of cancer.

**Radical prostatectomy:** an operation for prostate cancer in which the prostate, seminal vesicles and often a sample of some nearby lymph nodes are removed. It is an option in fit men and only when the urologist believes that the cancer is still confined to the prostate.

**Radiotherapy:** the use of radiation to kill cancer cells. With external-beam radiotherapy, the radiation is generated from an external source and focused onto the area of the prostate. See also 'Brachytherapy'.

**Recurrence:** when the cancer begins to grow again after a period of dormancy.

**Retrograde ejaculation:** this occurs following some types of surgery on the prostate. Instead of semen passing out through the penis during orgasm, it passes into the bladder, from which it passes out of the body when the man urinates.

**Risk factor:** a personal characteristic that increases the likelihood of getting a certain disease. The effect of a modifiable risk factor, such as a high-fat diet or smoking, can be reduced or overcome, in contrast to a non-modifiable risk factor such as belonging to an older age group or having a first-degree relative with the disease.

**Robotic radical prostatectomy:** a new way of performing a laparoscopic radical prostatectomy that employs the da Vinci® robot, which offers ten times magnification and three-dimensional vision, enabling very precise dissection of the prostate and preservation of the cavernous nerves.

**Scrotum:** the sac containing the testicles.

**Seminal vesicles:** storage vessels for sperm. They lie just behind the prostate and may be affected by prostate cancer as it spreads.

**Staging system:** a method used to assess and describe how far the cancer has spread. The tumour–nodes–metastases (TNM) system is commonly used in the UK.

**Testicles (or testes):** glands that make sperm and testosterone.

**Testosterone:** the androgen hormone responsible for the development of many male characteristics. It has a role in stimulating growth of prostate tissue, so some of the drugs for prostate cancer and BPH work by disrupting its production or effect.

**Tissue:** a collection of cells organized into a structure that performs a specific function.

**Transition zone:** the part of the prostate in which BPH usually develops.

**TRUS (transrectal ultrasonography):** an ultrasound method that allows the prostate to be seen. It involves inserting a lubricated ultrasound probe into the rectum, and is often used during brachytherapy and biopsy procedures so that the radiotherapist or doctor can see the exact position of the patient's prostate.

**TUIP (transurethral incision of the prostate):** an operation for BPH in which small nicks are made in the neck of the bladder and in the prostate to relieve the pressure on the urethra.

**TURP (transurethral resection of the prostate):** an operation for BPH in which the middle of the enlarged prostate is cut away piecemeal using an instrument inserted up through the penis.

**Ultrasound:** a method of forming images using high-frequency sound waves.

**Ureter:** one of two tubes that carry urine from the kidney to bladder.

**Urethra:** the tube that runs from the bladder to the tip of the penis, through which urine passes out from the body.

**Urodynamics:** a test to check how the bladder is functioning and whether the urine flow is blocked. It involves passing a fluid that will show up on scanning of the bladder (via a catheter) and then recording the movement of this fluid while the patient urinates.

**Uroflowmetry (urine flow test):** a test to measure the speed of urine output over time. It involves the patient passing urine into a specialized receptacle called a flow meter.

**Urologist:** a doctor who has specialized in disorders affecting the kidney, bladder and, in men, the prostate.

**Vas deferens:** a tube that carries sperm from the testis to the prostate gland.

**Vasopressin analogues:** drugs that may be prescribed to reduce the need to pass urine at night.

**Watchful waiting:** this involves delaying treatment until symptoms develop. The follow-up is less intense than that with active surveillance.

# Index

**A**
abiraterone  73
abscess, prostatic  102, 105
active surveillance  15, 43–44, 119
    in BPH  97
    locally advanced tumour  63
    long-term outcome  55, 57
acute bacterial prostatitis  100
adjuvant therapy  119
advanced cancer  69, 119
    bone-related problems  69–70
    locally advanced  36, 37, 63–66
        see also locally advanced tumour
    metastatic see metastases
    recurrent  59, 70–74, 125
age
    cancer risk  38–39
    PSA cut-off levels  13, 14
alcohol  51, 97
alfuzosin  107
alpha-blockers  87, 88, 105, 107, 119
    5-alpha reductase inhibitor with  88
alpharadin  74, 75
5-alpha reductase inhibitor  88, 107, 119
    alpha-blockers with  88
    in BPH  88
    dutasteride (Avodart)  41, 77, 88, 107
    finasteride (Proscar)  41, 88, 107
    side effects  88
alternative therapies  9, 97
American Cancer Society  115
anaesthesia, reaction and risks  91, 92
anal intercourse  101, 105
anatomy  3–4
androgens (testosterone and DHT)  64, 67, 88, 97, 121, 126
angiogenesis  37–38, 119
angiogenesis inhibitors  37–38, 73, 78

annual health checks  9–10
anti-androgens  53, 64, 109, 119
    hormone therapy in metastatic disease and  69
    monotherapy  66
    new drugs  78
    side effects  66
antibiotics  101, 105, 110
    before biopsy  28
    prostatic abscess  105
    prostatitis  101, 102, 103, 104
    resistance  104
anticholinergic drugs  89, 109, 120
anticoagulant drugs  8, 96
anti-inflammatory drugs  102–103, 110
antioxidants  7
anti-spasmodic drugs  109
anxiety  15, 17
    active surveillance and  44
apoptosis  79
aspirin  72
atorvastatin  41
Avastin (bevacizumab)  37–38
Avodart (dutasteride)  41, 77, 88, 107

## B

back passage see entries beginning rectal
bacterial prostatitis  100, 101
bad news  32
Bandolier  115
benign prostatic hyperplasia see BPH
benign tissue  120
Besttreatments (website)  115
bevacizumab (Avastin)  37–38
bicalutamide (Casodex)  64, 66, 109
biopsy  15, 17, 25, 37, 120
    cancer spread and  28–29
    consecutive/repeat  28
    transperineal  28–29
    ultrasound-guided  27–28
bisphosphonates  69–70
bladder  81

BPH tests  86–87
  urine volume in, ultrasound  86
Bladder and Bowel Foundation  115
bladder neck incision (TUIP)  92–93, 95, 127
bladder stones  82
blood, in urine  23, 28
blood pressure  22, 84
blood tests  23, 84–85
  in BPH  84–85
  ideal  18
  PSA see PSA (prostate-specific antigen) test
  in suspected prostate cancer  23
blueberry juice  8
bone metastases  37
  scans  29, 120
  treatment  69–70, 74
bone pain  69, 74
bone-related problems  69–70
bone scans  29
bowel problems  46, 64
BPH (benign prostatic hyperplasia)  4, 5, 81–97, 120
  active surveillance  97
  brachytherapy contraindication  48
  case study  96
  causes and mechanism  81, 82
  characteristics  81
  complications, risk  81–82
  development  4, 81, 82
  diagnosis  83–87
  early, moderate and advanced  82
  long-term outcome  95
  prevention  9, 97
  PSA levels  18, 19, 81, 85, 88
  radical prostatectomy for cancer with  54
  referrals  83
  risk factors  6, 81–82
  symptoms  25, 81, 83, 84
  treatment see treatment of BPH
  urine and blood tests  84–87
  urine flow tests  85–86
brachytherapy  47–49, 56, 58, 120

long-term outcomes 58
pros and cons 56
side effects and risks 48–49, 58
breast tenderness 66, 88
British Prostatitis Support Association 114
button electrodes 92

## C

cabazitaxel 72
cancer
    definition 13, 120
    prostate see prostate cancer
cardiovascular disease 6, 22, 72
Cardura XL 107
case study 40
    BPH 96
    HIFU (high-intensity focused ultrasound) 61
    hormone therapy and radiotherapy 65
    metastatic disease treatment 75
    palliative care 75
    prostatic intraepithelial neoplasia 40
    prostatitis 102
    recurrence of cancer 75
Casodex (bicalutamide) 64, 109
'castration' 67
catheters 49, 120
    in brachytherapy 48
    self-catheterization 54
cavernous nerves 49, 120
cell(s) 121
cell growth 35, 121
    prostate cancer 13, 14
central zone of prostate 3, 4
charities, cancer 32, 116
chemoprevention 40–41, 77
chemotherapy 70–72, 121
choline PET scans 31
chronic abacterial prostatitis 100
chronic bacterial prostatitis 100
chronic urinary retention 81
Cialis (tadalafil) 47, 53, 89, 110

colour Doppler  102
Combodart  88
complementary medicine  9, 97
conformal radiotherapy (CFRT)  46, 121
continence  121
    information sources  115
    see also incontinence
cranberry juice  7–8
creatinine  84
cryotherapy  59, 60
CT (computed tomography) scans  30, 31, 102, 121
CyberKnife  46, 121
Cyprostat (cyproterone acetate)  64, 69, 109
cystoscopy  121
Cystrin  109
cytoreduction see hormone therapy

## D

da Vinci® robot  51–52, 121
death rates  55, 57–58
Decapeptyl SR  108
de-differentiation  37
degarelix  108
denosumab  70
desmopressin  89, 110
Desmospray  89
Desmotabs  89
Detrunorm  109
Detrusitol XL (tolterodine)  89, 109
DHT (dihydrotestosterone)  88, 121
diabetes  23, 84
diagnosis
    BPH  83–87
    future improvements  77–78
    prostate cancer  13–15, 22–31, 77–78
    prostate cancer after TURP for BPH  92
    prostatitis  101–102
    suspected cancer  25–26
diazepam  105, 110
diclofenac (Voltarol)  102–103
diet  6–8, 51, 97, 113

differentiated tissue  121
digital rectal examination (DRE)  22, 23, 37, 121
    in BPH  83–84
diverticula  81, 82
docetaxel (Taxotere)  71–72
doctors
    specialists  21–22, 54, 70, 83, 127
    symptoms prompting visit to GP  10
    visiting your GP  9–10
Doralese  107
doxazosin  107
DRE see digital rectal examination (DRE)
drinking  7–8, 51, 97
driving after surgery  51
drug treatment  107–110
    BPH  87–89, 107, 109–110
    hormonal see hormone therapy
    see also specific drug groups
dutasteride (Avodart)  41, 77, 88, 107

# E

eating and drinking  6–8, 51, 97
ejaculation  5
    retrograde  90, 92, 94, 125
embarrassing problems, website  114
enzalutamide  73
erectile dysfunction see impotence
ethnicity  39
exercise  8, 51
    after radical prostatectomy  51
external-beam radiotherapy (EBRT)  45–46, 49, 56, 58, 65

# F

false-negative PSA results  17
false-positive PSA results  15, 17
family history  38–39
fats, dietary  6
fesoterodine  89, 109
finasteride (Proscar)  41, 88, 107
Firmagon  108
first-degree relatives  38–39, 122

'flare' phenomenon  64, 69
flavoxate  109
Flomaxtra (tamsulosin)  87, 88, 107
flutamide  109
function of prostate  4
future prospects, prostate cancer  77–79

## G

gene therapy  79
genetic mutations  41
genetics  38–39, 77
geographical variations, cancer  39–40
glands  122
Gleason score  25, 26, 31, 37, 122
    long-term outcome and  55
goserelin (Zoladex)  68–69, 108
grading of cancer  25–26, 37, 122
    imaging/tests for  26–31
GreenLight laser prostatectomy  95, 122
growth factors  73
    inhibitors  37–38, 73, 78

## H

health checks  9–10, 22, 84
health insurance  21
Health of Men  113
Health Press  114
healthy eating  6–8
herbal remedies  8, 9, 97
HIFU (high-intensity focused ultrasound)  58, 59–60, 122
    case study  61
    probe  59
    side effects  59–60
Holmium laser enucleation of the prostate (HoLEP)  95–96, 122
hormone(s)  122
    female (oestrogens)  72, 97
    male (androgens)  64, 67, 88, 97, 121, 126
hormone-relapsed prostate cancer  70–72, 75, 122
hormone therapy  107–109, 123
    aims  64
    for BPH  88, 107

case study 65
  effectiveness 64–65
  effect on cancer cells 71
  intermittent 65
  for locally advanced cancer 63–66
  for metastatic cancer 68–69
  radical prostatectomy after 66
  radiotherapy after 46, 48, 58, 66
  for recurrent cancer 72
  side effects 64, 66, 69, 72, 88, 107, 108
  with surgery 53, 66
hot flushes 64, 68
'hot spots,' bone scans 29
hypertension 84
Hytrin 107

# I

imaging tests 29–31
immunotherapy 73, 78–79, 123
impotence 123
  after hormone therapy 64, 69
  after radiotherapy 46–47, 49, 54
  after surgery 49, 53, 54, 67, 94
  after treatment for BPH 88, 90, 91, 94
  after TURP 90, 91
  information sources 115
  treatment of 47, 53, 115
  zinc deficiency 7
incontinence 53, 54, 115
  after radical prostatectomy 5, 53
  after TURP 91
individualised cancer tests 78
indoramin 107
infections 7–8, 28, 100, 102, 105
infertility 53, 67–68, 90
inflammation, prostate see prostatitis
inflammatory prostatitis, asymptomatic 100
information sources 19, 32, 113–116
informed choice, PSA testing 19

intensity-modulated radiotherapy (IMRT)  46, 123
intermittent hormone therapy  65
internet sites  113–116

**K**

keyhole surgery  50–52
kidney disease  81, 84

**L**

laparoscopic prostatectomy  50–52
laser prostatectomy  95–96, 122, 123
latent prostate cancer  25
leucopenia  72
leuprorelin  108
Levitra (vardenafil)  47, 53
LHRH analogues  63, 64, 108, 123
    before brachytherapy  48
    intermittent hormone therapy  65
    in locally advanced cancer  63, 64, 65
    maximal androgen blockade  69
    in metastatic disease  68–69
LHRH antagonists  63–64, 69, 108, 123
life expectancy  1, 38
lifestyle
    after surgery  51
    preventative  6, 39–41, 44, 102, 105, 113–114
lifetime risk, prostate cancer  38
Lipitor (atorvastatin)  41
liver function  64, 66
locally advanced tumour  35, 36, 37
    cryotherapy  59
    description  63
    treatment  63–66
long-term outcome  55–58
    after surgery for BPH  95
LUTS (lower urinary tract symptoms)  83, 84, 123
lycopene  7
lymphatic system  36, 123
lymph nodes  26, 49, 123
Lyrinel XL  109

## M

Macmillan Cancer nurses  74
Macmillan Cancer Support  114, 116
malignant disease see cancer; prostate cancer
Marie Curie Cancer Care  116
Marie Curie nurses  74
maximal androgen blockade  67, 69, 124
Men's Health Forum  113
metabolic rate  8
metastases  36, 37, 124
    bone see bone metastases
    diagnosis  29–31
    mechanism  36, 37
    not caused by biopsy  29
    staging of prostate cancer  26
    treatment  74
metastatic disease treatment  66–70
    case study  75
    hormone therapy  68–69
    maximal androgen blockade  67, 69
    orchidectomy  67–68
minerals  7
MRI (magnetic resonance imaging)  30, 43, 124
multi-leaf collimator  46

## N

National Cancer Institute (USA)  115
nerve-sparing approach, prostatectomy  49
nervous system disorders  83–84
NHS Direct  113
NICE  114, 115
nocturia  24, 89
nurse specialists  32

## O

obesity  6, 22
oestrogens  72, 97
oncologists  70, 124
open prostatectomy  93–95, 124
orchidectomy  67–68, 124
osteoporosis  68

over-the-counter tests  17
overweight  6, 8
oxidative stress  101
oxybutynin  109

**P**
painkillers  74
palliative care  73–74, 124
    case study  75
Partin's tables  31
pathologists  27, 124
PCA3 (prostate cancer antigen 3)  77, 78, 124
pelvic pain syndrome (prostatodynia)  100, 105, 125
perineum  5, 124
peripheral zone of the prostate  3, 4, 125
PET scans  31
physical examination  23, 83–84
phytotherapy  8, 9, 97, 125
PIN see prostatic intraepithelial neoplasia (PIN)
PLACE mnemonic  6
plant extracts (phytotherapy)  8, 9, 97, 125
plasma button prostatectomy  125
'positive margin'  53
pregnant women  88
prevention  5–10
    BPH  9, 97
    chemoprevention  40–41
    prostate cancer  5–10, 39–41, 77
    prostatitis  105
primary tumours  36
prognosis  66
Prolaris test  78
propiverine  109
Proscar (finasteride)  41, 88, 107
prostaglandin injections  53
Prostap  108
prostate
    anatomy  3–4
    increasing interest in  1
    inflammation see prostatitis
prostate cancer  5

    case studies see case study
    cell growth  13, 14
    clinically insignificant, overdetection  15
    development and causes  35–38, 40
    diagnosis  13–15, 22–31, 77–78
    grading  25–26, 37, 122
    ideal blood test  18
    prevention  5–10, 39–41, 77
    PSA in see PSA (prostate-specific antigen)
    recurrence see recurrence of prostate cancer
    risk factors  38–39
    screening  9–10, 15, 77
    spread, Partin's tables  31
    support groups  114
    treatment see treatment of prostate cancer
    vitamin D protective action  7
    'window of curability'  9, 10
prostate cancer antigen 3 (PCA3)  77, 78, 124
Prostate Cancer UK  74, 114
prostatectomy
    laser  95–96, 122, 123
    open  49–50, 93–95, 124
    radical see radical prostatectomy
prostate disorders, information sources  114
prostate-specific antigen see PSA
prostatic abscess  102, 105
prostatic diseases  4–5
prostatic intraepithelial neoplasia (PIN)  13, 36–37, 125
    case study  40
prostatic massage  101, 103
prostatic secretions, sample  101
prostatitis  5, 99–105
    case study  102
    classification  100
    diagnostic tests  101–102
    prevention  105
    PSA levels  18, 19, 102
    recurrent  102, 103–104
    risk factors  101, 105
    symptoms  99
    treatment  102–104

prostatodynia  100, 105, 125
Provenge (sipuleucel-T)  73, 79
PSA (prostate-specific antigen)  4, 9–10, 15, 125
    after radical prostatectomy  52, 53
    bound vs free  15
    in BPH  18, 19, 81, 85, 88
    cut-off with age  13, 15
    elevated
        conditions with  15, 18, 19
        next steps  21–32
    free, reduced in prostate cancer  15, 21
    high cut-off  17
    kinetics  9
    leakage mechanism  13, 14
    normal  13, 14, 18
    in pre-cancer stage  13
    in prostate cancer  10, 13–19, 52–53
        aggressive tumours  17
        recurrence  58
        scan indications  31
        small/early cancers  13, 17, 37
    in prostatitis  18, 19, 102
    sudden rise in  10, 13–14, 17, 19
PSA test  9–10, 13–19
    active surveillance and  44
    anxiety before results  15, 17
    basis  13–15
    false-negative results  18
    false-positive results  15, 17
    'ideal' blood test relationship  18
    informed choice  19
    issues involving  15–17
    pros and cons  16
    regular  9–10, 17, 19, 43

## Q

quality of life  15, 43–44
questionnaire, urinary symptoms  24, 25

## R

race 39
radical prostatectomy 49–58, 55–58, 66, 125
    after hormone therapy 66
    further treatment after 53
    indications for 49
    laparoscopic and robotic 50–52
    long-term outcome 55, 56, 57–58
    'open' procedure 49–50, 93–95, 124
    postoperative care 51
    pros and cons 56
    PSA levels after 52, 53
    questions to ask before 54
    radiotherapy choice vs 55, 56
    recurrence rate 57–58
    robotic 51–52, 126
    side effects and risks 53–54
radionuclide scans 29
radiotherapy 45–49, 55, 56, 58, 125
    after hormone therapy 46, 48, 58, 66
    conformal (CFRT) 46
    external-beam (EBRT) 45–46
    intensity-modulated (IMRT) 46
    long-term outcomes 58
    or radical prostatectomy, choice? 55, 56
    palliative 74
    pros and cons 55, 56
    side effects and risks 46–47, 48–49, 53–54, 58
    see also brachytherapy
rectal cancer 46
rectal examination see digital rectal examination (DRE)
recurrence of prostate cancer 70–74, 125
    after brachytherapy 49, 58
    after radical prostatectomy 57–58
    after radiotherapy 58, 59
    case study 75
    treatment see treatment of recurrence of cancer
REDUCE study 41, 77
referral 21, 83
Regurin 109
retrograde ejaculation 90, 92, 94, 125

risk factors  126
    for BPH  6, 81–82
    for complications of BPH  81–82
    for prostate cancer  38–39
    for prostatitis  101, 105
robotic radical prostatectomy  50, 51–52, 126

## S

SAGA Health  114
saw palmetto  8, 97
screening for cancer  9–10, 13–19
    targeted  77
scrotum  67, 126
secondary tumours  36
    see also metastases
selenium  7
self-catheterization  54
self-help  5–9
semen  4, 5
    formation and ejaculation  4, 5
seminal vesicles  4, 49, 50, 126
septicaemia  28
sexual activity after surgery  51, 52
sexual dysfunction see impotence
sexually transmitted infections  100, 105
side effects
    alpha-blockers  87, 107
    5-alpha reductase inhibitors  88
    anti-androgens  66
    brachytherapy  48–49, 58
    chemotherapy  70, 72
    HIFU  59–60
    hormone therapy  64, 66, 69, 72, 88, 107, 108
    orchidectomy  67–68
    prostatectomy  53–54, 94
    radiotherapy  46–47, 48–49, 53–54, 58
    TUIP  92–93
    TURP  90–91
sildenafil (Viagra)  47, 53
sipuleucel-T (Provenge)  73, 79
size of prostate  3

smoking  8
social support  32, 116
solifenacin (Vesicare)  89, 109
specialist nurse  32
sperm banking  53
spread of cancer  35, 36, 37
    Partin's tables  31
stages of cancer  36–38
staging system for cancer  26, 126
    imaging/tests for  26–31
statins  41, 77
steroids  72
strontium  74
sunlight  39–40
support groups  44, 114, 116
surgery
    for BPH  89–96
    for cancer  49–58, 55–58, 66, 67–68
    see also radical prostatectomy
survival rates
    radical prostatectomy  55, 57–58
    radiotherapy  58
'Survivorship' programme  74
symptoms prompting GP visits  10

## T

tadalafil (Cialis)  47, 53, 89, 110
tamsulosin (Flomaxtra)  87, 88, 107
targeted screening  77
Taxotere (docetaxel)  71–72
template transperineal biopsy  28–29
terazosin  107
terminal care  73–74
testicles  126
testosterone  3, 64, 67, 97, 126
    orchidectomy effect  67–68
tests  22–25
    blood see blood tests
    PSA see PSA (prostate-specific antigen) test
    urine  23, 84, 101
tissue  126

TNM (tumour–nodes–metastases) staging system  26, 66
    T3-N1-M1  66
tolterodine (Detrusitol XL)  89, 109
tomatoes  7
Toviaz  89, 109
trace elements  7
transition zone of the prostate  3, 4, 126
transperineal biopsy  28–29
transrectal ultrasonography (TRUS)  26, 102, 126
    BPH  87
    prostatitis  102
transurethral incision of the prostate (TUIP)  92–93, 95, 127
transurethral resection of the prostate (TURP)  48, 89–92, 95, 127
    side effects  90–91
treatment of BPH  87–97
    active surveillance  97
    drugs  87–89, 107, 109–110
    minimally invasive techniques  95–96
    phytotherapy  97
    surgery  54, 89–96
treatment of prostate cancer
    active surveillance  43–44, 55, 63
    chemotherapy  70–72
    combination therapies  46, 48, 53, 58, 66
    experimental  37–38, 59–62, 73, 78–79
    hormone therapy  63–66, 68–69, 72, 108–109
    information sources  115
    making a decision  55, 115
    monitoring  52–53
    new, future prospects  78–79
    radiotherapy see radiotherapy
    recurrent disease  59, 70–74
    surgery  49–58, 55–58, 66, 67–68
        see also prostatectomy
    survival rates  55, 57–58
treatment of prostatitis  102–104
treatment of recurrence of cancer  70–74
    chemotherapy  70–72
    cryotherapy  59
    hormone therapy  72
    new treatments  73

palliative 73–74
triptorelin 108
trospium 109
TRUS (transrectal ultrasonography) 26, 87, 102, 126
TUIP (transurethral incision of the prostate) 92–93, 95, 127
tumour nodule 35
tumours 35, 36
    benign vs malignant 36
    see also prostate cancer
TURP see transurethral resection of the prostate (TURP)

## U

UK Prostate Link 114
ultrasound 127
    biopsy guided by 27–28
    for BPH diagnosis 86–87
    for diagnosis 26, 86–87, 102
    HIFU see HIFU (high-intensity focused ultrasound)
    urine volume in bladder 86
urethra 3, 4, 81, 82, 91, 127
urethral stricture 91
urinary incontinence see incontinence
urinary symptoms 10, 24
    after radical prostatectomy 53–54
    in BPH 81, 83, 84
    prostatitis 99
    questionnaire 24, 25
    radiotherapy side effects 46
    see also incontinence
urinary tract
    infections 7–8, 28, 100, 102, 105
    symptoms of disease see urinary symptoms
    treatment of symptoms 89, 109–110
urine
    blood in 23, 28
    tests on 23, 84, 101
urine flow tests 85–86, 127
Uripass 109
urodynamics 86–87, 127
uroflowmetry 85–86, 127

urologists 21–22, 83, 127
    questions to ask, before prostatectomy 54
urology nurse specialist 32

## V
vardenafil (Levitra) 47, 53
vas deferens 127
vasopressin analogues 89, 110, 127
Vesicare (solifenacin) 89, 109
Viagra (sildenafil) 47, 53
vitamin D 7, 39–40
vitamin E 7
Volterol (diclofenac) 102–103

## W
warfarin 8
watchful waiting 44, 127
    see also active surveillance
websites, information 113, 114, 115, 116
weight 6, 8
'window of curability' 9, 10
working after surgery 51

## X
Xatral XL 107

## Z
zinc 7
Zoladex (goserelin) 68–69, 108
Zometa (zoledronic acid) 69